Springer Tracts in Modern Physics 105

Editor: G. Höhler
Associate Editor: E. A. Niekisch

Editorial Board: S. Flügge H. Haken J. Hamilton
H. Lehmann W. Paul

Springer Tracts in Modern Physics

---

* denotes a volume which contains a Classified Index starting from Volume 36.

Hermann Kolanoski

Two-Photon Physics at e⁺e⁻ Storage Rings

With 123 Figures

Springer-Verlag Berlin Heidelberg GmbH 1984

Dr. Hermann Kolanoski

Physikalisches Institut der Universität Bonn, Nußallee 12,
D-5300 Bonn 1, Fed. Rep. of Germany

Manuscripts for publication should be addressed to:

Gerhard Höhler

Institut für Theoretische Kernphysik der Universität Karlsruhe
Postfach 6380, D-7500 Karlsruhe 1, Fed. Rep. of Germany

*Proofs and all correspondence concerning papers in the process of publication
should be addressed to:*

Ernst A. Niekisch

Haubourdinstrasse 6, D-5170 Jülich 1, Fed. Rep. of Germany

ISBN 978-3-662-15732-9   ISBN 978-3-540-39033-6 (eBook)
DOI 10.1007/978-3-540-39033-6

Library of Congress Cataloging in Publication Data. Kolanoski, Hermann, 1945– Two-photon physics at e+ e⁻
storage rings. (Springer tracts in modern physics; 105) Bibliography: p. Includes index. 1. Photonphoton
interactions. 2. Hadrons-Scattering. 3. Quantum chromodynamics. 4. Storage rings. I. Title. II. Series.
QC1.S797   vol. 105   539 s   84-13894 [QC794.8.P4]   [539.7'217]

Preface

The investigation of two-photon interactions is a relatively young field of experimental particle physics and is closely connected to the advent of high energy electron-positron storage rings. In a storage ring the inter-actions between the clouds of virtual photons accompanying both beams can be used to study photon-photon reactions over a wide kinematical range. The majority of experimental results in this field has been ob-tained only in the last five years, in particular at the storage ring PETRA in Hamburg. In two-photon reactions with leptonic final states quantum electrodynamics can be tested up to the fourth order of the fine structure constant α. In the last few years, the experimental activities have concen-trated on studies of two-photon production of hadrons and results have been obtained on the two-photon coupling of meson resonances, on the production of hadrons with large transverse momenta and on the struc-ture of the photon as measured in deep-inelastic electron-photon scattering. The simple and controllable initial state in two-photon inter-actions allows hadron dynamics to be probed cleanly and allows tests of quantum chromodynamics, the theory describing strong interactions among hadrons.

This report reviews the experimental status of two-photon physics and discusses to what extent the theoretical predictions for two-photon re-actions have been corroborated by experiment.

Bonn, January 1984   H.Kolanoski

Acknowledgements

This review of two-photon physics has grown out of research done together with the TASSO collaboration at the storage ring PETRA in Hamburg. I would like to take this opportunity to thank all members of the collaboration, in particular the spokesman Dr. G. Wolf, for a very satisfactory cooperation.

Within the TASSO collaboration, the group from the University of Bonn provided a particularly pleasant working atmosphere - friendly, stimulating, encouraging. The present work is heavily based on the continuous support I received from the members or former members of the group: Drs. H. Boerner, S. Cooper, H.M. Fischer, H. Hartmann, W. Hillen, E. Hilger, G. Knop, L. Köpke, H. Kück, P. Leu, B. Löhr, R. Wedemeyer, N. Wermes and M. Wollstadt. For particularly valuable support during the preparation of the manuscript I want to thank Profs. E. Hilger and R. Wedemeyer. For the encouragement to write this report and for providing the conditions which made it possible, I thank Prof. G. Knop. To Prof. W. Paul I am grateful for valuable comments and for supporting the publication of this report.

I have benefited from discussions and comments from many colleagues. In particular I want to mention: Drs. Ch. Berger, D. Cords, J. Field, G. von Gehlen, F. Gutbrod, J. Kühn, J. Olsson, B Schrempp, F. Schrempp and W. Wagner. I am grateful to Drs. B. Foster and D. Pandoulas for carefully reading the final version of the manuscript.

The work on the review originally started from lectures which I gave at the XXII. Cracow School of Theoretical Physics at Zakopane in June 1982. I would like to thank the organizers of this school, in particular Profs. A. Biaƚas and K. Zalewski, for the kind invitation and for the hospitality I received during my stay in Poland.

Finally, I would like to thank my wife Ria and my daughters Saskia, Julia and Martina for their patience and for allowing me the time and space within the family to do this work.

Contents

1. Introduction

According to Maxwell's classical linear equations, electromagnetic waves cross each other without disturbance. However, the quantized theory of electromagnetism, quantum electrodynamics, predicts that interactions occur between the field quanta, the photons, by way of quantum fluctuations of the vacuum. For small photon energies, as in the optical range, the cross sections are extremely small. Therefore, photon-photon scattering has only been observed at large energies, i.e. above the threshold for the production of electron-positron pairs.

Very intense photon sources are provided by storage rings for highly energetic electron and positron beams. These rings were built to investigate the annihilation of electrons and positrons. In lowest order of the electromagnetic coupling constant α, $O(\alpha^2)$, that process can be viewed as the annihilation of e^+e^- into a virtual (timelike) photon which then couples to a final state X of leptons or hadrons (Fig.1.1a).

The two-photon production of a final state X,

$$e^+e^- \rightarrow e^+e^- \; X,$$

as shown in Fig.1.1b is of order α^4. Each of the two leptons radiates a photon and the two photons produce the final state X. In this case the produced system X has even charge conjugation, whereas in the one-photon annihilation process the final state has odd charge conjugation. Involving two electromagnetic currents, two-photon reactions exhibit a rich dynamical structure which supplies additional and often complementary physics information to that obtained from one-photon annihilation processes.

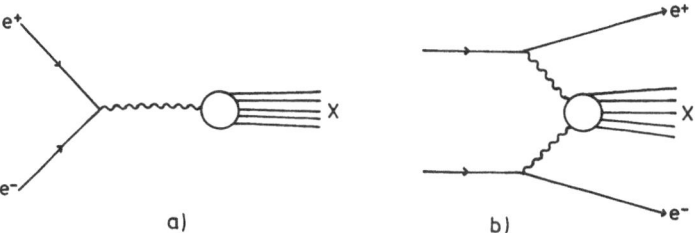

<u>Fig.1.1.</u>   Production of a final state X in e^+e^- collisions: a) by one photon and b) by two photons.

1.1 Physics with Two Photons

The cross sections arising from diagram b) increase with the beam energy E like $(\log E/m_e)^2$, while those from a) decrease at least like $1/E^2$. Thus it turns out that the two-photon process, although of order α^4, already dominates over the order α^2 annihilation process at beam energies of a few GeV. However, most of the events from the two-photon process have low invariant masses of the produced system due to the typical bremsstrahlung spectrum, $\sim 1/E_\gamma$, of the radiated photons.

The photon propagators in diagram b) cause the bulk of the photons to be radiated nearly on-mass-shell at small angles relative to the beam. This effectively provides two colliding beams of quasi-real photons, with a small beam divergence of order m_e/E. Hence we talk about photon-photon scattering, regarding the electron and positron beams just as sources for the "photon beams". The high energy machines available today, like PETRA and PEP, provide powerful photon beams with high two-photon luminosities[1]. Integrating over invariant photon-photon masses above the pion pair threshold, the two-photon luminosities at beam energies around 15 GeV are only an order of magnitude smaller than the e^+e^- luminosities. Whereas for a given beam energy the e^+e^- kinematics of an annihilation process is fixed, the continuous spectra of the photon beams allow simultaneous measurements at different $\gamma\gamma$ invariant masses and for different momentum transfers and polarisations of the photons.

Quantum electrodynamics (QED) can be tested up to order α^4 by measuring two-photon production of lepton pairs. The investigation of two-photon production of hadronic final states provides the possibility of probing hadron dynamics with a simple, calculable initial state. Our main interest lies in the study of the coupling of the photon to hadrons, which in the framework of the quark model means the coupling to quarks as the constituents of hadrons. In the resonance region the quark model of bound quark-antiquark systems can be tested by measuring the two-photon resonance couplings. A large fraction of the experimental results obtained in the last years deals with these couplings. Enough information is now available to allow meaningful tests of SU(3) symmetry and of models with multiquark or gluonium states. The possible existence of bound states of gluons (gluonia) has been discussed extensively in recent years and the measurements of the $\gamma\gamma$ widths of resonances provide important inputs to this debate.

The study of hard scattering processes in two-photon reactions was made possible when beam energies up to about 20 GeV became available at the e^+e^- storage rings PETRA at Hamburg and PEP at Stanford. The investigation of the production of particles with high transverse momenta, jet production, and scattering of highly virtual photons (deep-inelastic scat-

---

[1]   The luminosity L of interacting beams determines the rate $\dot{N}$ at which a reaction with a cross section σ occurs according to the equation: $\dot{N}=L\cdot\sigma$.

tering), allows tests of quantum chromodynamics (QCD), the widely
accepted theory of strong interactions.

Two-photon production of hadrons proceeds in lowest order as in
Fig.1.2a where the two photons have a pointlike QED coupling to a quark
pair and the quarks subsequently fragment into hadrons. However, this
pointlike behaviour of the photon is observable only at high momentum
transfer, in interactions at small distances. We know that the vector
meson dominance (VMD) model works rather well in most processes involving real or quasi-real photons. In the VMD picture the photons transform
into virtual vector mesons (ρ, ω, φ, ...) which then interact strongly with
hadrons (Fig.1.2b). According to this model photon-photon scattering
should exhibit the characteristic features of hadronic interactions observed in hadron-hadron scattering.

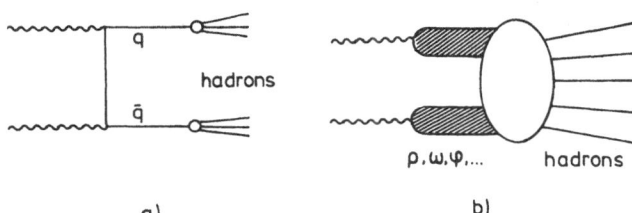

a)   b)

__Fig.1.2.__   a) Pointlike and b) hadron-like two-photon interactions.

Since the "hadron-like" and the "pointlike" pieces arise from the same
photon, we expect a smooth transition between the two pictures. The
scale-free coupling in Fig.1.2a is reached only asymptotically for high
transverse momenta of the quarks or for a high momentum transfer of at
least one of the photons to the quarks. As the momentum transfer becomes smaller, the interaction is spread over larger distances and longer
times, so that the quarks can interact via gluon exchange and form bound
states carrying the quantum numbers of the photon (vector mesons) as
sketched in Fig.1.3.

In two-photon reactions this dual nature of the photon can be
investigated: In the regime of low four-momentum transfer Q^2 of the
photons and low transverse momenta of the produced hadrons the
hadronic character of the photon will dominate (total cross section of
quasi-real photons). At large Q^2 and/or high transverse momenta of the
hadrons the elementary, pointlike nature of the photon is emphasized
(high-p_T jets, structure functions).

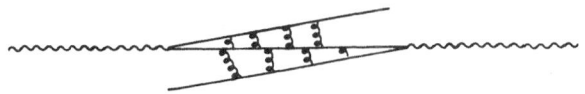

__Fig.1.3.__   Sketch of $\gamma\gamma$ coupling to quarks with low transverse momentum.

1.2   The Historical Development of Two-Photon Physics

The first calculations of two-photon reactions, specifically elastic photon-photon scattering, were published about 50 years ago /1/. Much later, in 1960, the first theoretical papers relevant to two-photon physics at e^+e^- storage rings appeared. F.Low /2/ suggested measuring the two-photon coupling of the π° at storage rings via

$e^+e^- \rightarrow e^+e^-\pi^\circ$

and F.Calogero and C.Zemach /3/ calculated the process

$e^+e^- \rightarrow e^+e^-\pi^+\pi^-$.

A large number of papers appeared at the end of the sixties and early seventies, when the storage rings in Novosibirsk, Frascati, Orsay, Stanford and Hamburg became available. Most of the theoretical work on low energy and resonance physics with two photons stems from that time and excellent reviews are available /4/. For a rather complete bibliography of the theoretical work from that period see under ref. /4/ the review article of Budnev et al..

At about the same time the first experimental results on two-photon QED reactions were obtained in Novosibirsk and Frascati /5, 6/. With the discovery of the charm quark in one-photon annihilation the general interest turned in this direction and the experimental and theoretical activities in two-photon physics were reduced to a minimum.

In view of the high energies which became available at PETRA and PEP, theorists realized that two-photon physics provides an excellent opportunity to test QCD by studying large-p_T hadron production and jet formation /7/ and by measuring the structure function of the photon /8/. The experimental activity was resumed with the measurement of the two-photon production of the η' at the storage ring SPEAR in 1979. In the following years much experimental information was collected on resonance couplings, total cross sections and hadron pair production. Experiments at PETRA started to look for hard scattering processes such as two-photon jet production and particle production at large transverse momenta. Much effort has also been devoted to the measurements of the structure function of the photon and the first attempts to determine from these measurements the QCD scale parameter Λ.

At the moment we seem to be in a time of fruitful activity and continuous progress in two-photon physics. Many new experimental results are expected in the near future.

1.3   Organization of the Review

This article reviews the experimental achievements of two-photon physics at e^+e^- storage rings in the last years. In general, we want to concentrate on published results and neglect preliminary data unless they contribute new aspects to a topic. The field is still young enough so that an almost complete coverage of the experimental results seems possible. Theoretical ideas and models are discussed in so far as they are relevant for the understanding of the experimental results. No attempt is made to review the whole theoretical work accumulated over the years, references are mainly given to some basic work and to reviews. In preparing this report, the conference talks on two-photon physics in recent years /9/ and particularly the proceedings of the two-photon meetings /10/ were of substantial help.

The article is organized as follows: In Chap.2 the two-photon kinematics and the general form of the cross section are introduced. Chapter 3 gives a short summary of the characteristics of the detectors mentioned later. The review of experimental results starts in Chap.4 with a brief overview of the QED results. Two-photon meson and baryon pair production is discussed in Chap.5. Much room is devoted in Chap.6 to the review of the experimental results on two-photon couplings of resonances and the discussion of their theoretical implications. The following three chapters deal with measurements of the two-photon total cross section: in Chap.7 the total cross section measurements at low Q^2 are discussed, Chap.8 focuses on the production of hadrons with high transverse momenta and in Chap.9 the case is studied where at least one of the photons is highly virtual, allowing the measurement of the structure of the photon. Finally, Chap.10 summarizes the experimental results and lists the problems and open questions which might be tackled in the near future.

2. Kinematics and Cross Section of Two-Photon Reactions

2.1 The Kinematics of the Reaction $e^+e^- \to e^+e^-X$

   Two-photon scattering in e^+e^- storage rings can be observed in the reaction

$$e^+e^- \to e^+e^-X \qquad\qquad (2.01)$$

according to the lowest order diagram in Fig.2 1a: an electron and a
positron radiate photons, which produce a particle system X with even
C parity. Though this is not the only diagram leading to the reaction
(2.01), in most cases it is the dominant one. However, in certain
kinematical regions the "virtual bremsstrahlung processes" with negative
C parity of the final state X (Fig.2.1b,c) can also give a sizeable contribution /11/.
   Due to the photon propagators in diagram a) the photons in this reaction are emitted predominantly at small angles, of order m_e/E, with respect to the beam. That leads to small (space-like) momentum transfers

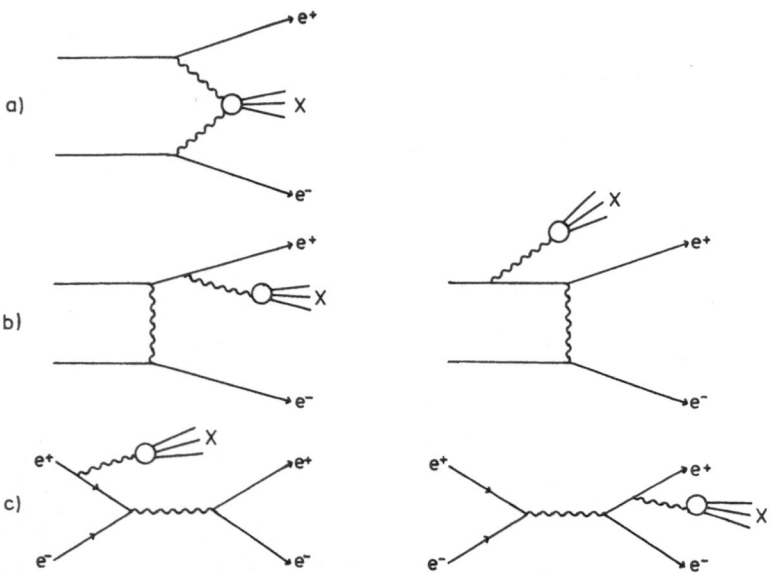

Fig.2.1.   Diagrams contributing to the reaction $e^+e^- \to e^+e^-X$: In diagram a) X is a C-even state, in diagrams b) and c) X is a C-odd state.

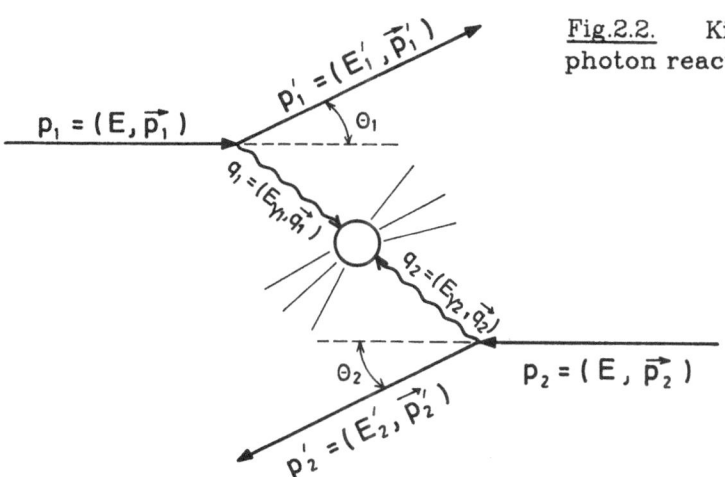

Fig.2.2.   Kinematics of the two-photon reaction $e^+e^- \to e^+e^-X$.

to the system X. The invariant mass of the system X peaks at small values because the photon energies follow roughly the characteristic bremsstrahlung spectrum $\sim 1/E_\gamma$.

For the inclusive detection of X the kinematics of reaction (2.01) is completely determined by the four–momenta of the incoming and of the scattered electron and positron (p_1, p_2 and p_1', p_2', respectively; see Fig.2.2). For unpolarized lepton beams (i.e. if there is no overall azimuthal dependence) five variables are needed to determine the $\gamma\gamma$ system at a given beam energy E. The following variables are defined in the laboratory system:

—   the energies E_1', E_2' of the scattered leptons,

—   their angles Θ_1, Θ_2 with respect to the beam axis,

—   the angle Φ between the two lepton scattering planes.

The kinematical variables of the $\gamma\gamma$ system can be expressed in terms of the variables of the scattered leptons. The energies and normalized energies of the photons are:

$$E_{\gamma 1} = E - E_1' \quad \text{and} \quad E_{\gamma 2} = E - E_2',$$
$$\omega_1 = E_{\gamma 1}/E \quad \text{and} \quad \omega_2 = E_{\gamma 2}/E. \tag{2.02}$$

The squared masses of the (space–like) photons are:

$$q_1{}^2 = -Q_1{}^2 = (p_1 - p_1')^2 = 2m_e{}^2 - 2EE_1'(1 - \sqrt{1 - (m_e/E)^2} \cdot \sqrt{1 - (m_e/E_1)^2} \cdot \cos\Theta_1)$$

$$q_2{}^2 = -Q_2{}^2 = (p_2 - p_2')^2 = 2m_e{}^2 - 2EE_2'(1 - \sqrt{1 - (m_e/E)^2} \cdot \sqrt{1 - (m_e/E_2)^2} \cdot \cos\Theta_2). \tag{2.03}$$

For $\Theta_1 \gg m_e/E$ (2.03) reduces to

$$q_1{}^2 \approx -2EE_1'(1 - \cos\Theta_1). \tag{2.04}$$

Since the scattering angles Θ_i are predominantly very small, the electron mass terms cannot always be neglected. In particular, one obtains for $\Theta_i=0$ the minimum Q_i^2:

$$Q_{i\,min}^2 = m_e^2 \cdot \omega_i/(1-\omega_i) + O(m_e^4). \tag{2.05}$$

The invariant mass of the $\gamma\gamma$ system is (neglecting terms of order m_e^2):

$$W_{\gamma\gamma}^2 = (q_1+q_2)^2 = 4 \cdot E_{\gamma 1} \cdot E_{\gamma 2} - 2 \cdot E_1' E_2' \cdot (1-\cos\Theta_1 \cdot \cos\Theta_2 - \sin\Theta_1 \cdot \sin\Theta_2 \cdot \cos\Phi) \tag{2.06}$$

which for small Q_1^2 and Q_2^2 can be approximated by:

$$W_{\gamma\gamma}^2 \approx 4 \cdot E_{\gamma 1} \cdot E_{\gamma 2}. \tag{2.07}$$

For the definition of the photon structure functions (Chap.9) the following invariant variables are useful:

$$
\begin{aligned}
y_1 &= q_1 \cdot q_2/p_1 \cdot q_2 \quad \text{and} \quad x_1 = Q_1^2/2q_1 \cdot q_2 \\
y_2 &= q_1 \cdot q_2/p_2 \cdot q_1 \quad\quad\quad\; x_2 = Q_2^2/2q_1 \cdot q_2.
\end{aligned} \tag{2.08}
$$

2.2 The Hélicity Structure of the Two-Photon Cross Section

The transition matrix element for the reaction $e^+e^- \to e^+e^-X$ as given by diagram a) in Fig.2.1 can be factorized into three terms: one for the electron-photon vertex, one for the positron-photon vertex and one for the coupling of the two photons to the system X. This last term is the unknown piece which has to be measured. The $ee\gamma$ vertices determine the fluxes and polarization states of the two photons. The factorization of the total matrix element into a part which generates photons and another part which describes the scattering of these photons, allows one to talk about (virtual) two-photon scattering at e^+e^- machines. The transition matrix element T for the $\gamma\gamma X$ vertex depends on the polarization states of the two photons as determined by the polarization vectors ε_1 and ε_2:

$$\langle X|T|\gamma\gamma\rangle = \varepsilon_1^\mu \cdot \varepsilon_2^\nu \cdot A_{\mu\nu}^X, \tag{2.09}$$

where $A_{\mu\nu}^X$ is a tensor which contains the helicity structure of the transition. Since conservation of the electromagnetic current imposes the constraints

$$q_{i\mu} \cdot \varepsilon_i^\mu = 0 \quad (i=1,2) \tag{2.10}$$

the photon polarization can be expressed by a 3-vector $\vec{\varepsilon}_i$. Consequently, any mixture of pure polarization states of a photon can be given by a 3×3 density matrix $\rho_i^{\mu\nu}$ (i=1,2 for the two photons). In the diagram of Fig.2.1a the photons are produced with a linear polarization with respect to the lepton scattering planes. As can be derived from the $ee\gamma$ vertices the polarization of each photon is determined by two vectors, one in the lepton scattering plane and one perpendicular to it. The vector in the plane has both transverse and longitudinal components. The corresponding (unnormalized) density matrices are /12, 13/:

$$\rho_1{}^{\mu\nu} = -[q_1{}^2\ g_{\mu\nu} + 2\cdot(p_1{}'^{\mu}\ p_1{}^{\nu} + p_1{}^{\mu}\ p_1{}'^{\nu})]/q_1{}^2. \tag{2.11}$$

In terms of these density matrices the differential cross section for reaction (2.01) can be written /12/:

$$d\sigma(e^+e^-\to e^+e^-X) = \frac{\alpha^2}{64\pi^4\ E^2\ q_1{}^2q_2{}^2}\cdot \rho_1{}^{\mu\nu}\ \rho_2{}^{\mu'\nu'}\ M^X{}_{\mu\nu\mu'\nu'}\cdot \frac{d^3p'_1\ d^3p'_2}{E_1{}'\cdot E_2{}'}. \tag{2.12}$$

The tensor $M^X{}_{\mu\nu\mu'\nu'}$, which describes the production of the final state X, is given by:

$$M^X{}_{\mu\nu\mu'\nu'} = 8\pi^2\int A^X{}_{\mu\nu}{}^*A^X{}_{\mu'\nu'}\delta(q_1+q_2-P_f)d\Gamma_f. \tag{2.13}$$

P_f is the four-momentum of the final state and the integral goes over the final state phase space, denoted by Γ_f. In general, $M^X{}_{\mu\nu\mu'\nu'}$ cannot be calculated rigorously for hadronic final states. However, by Lorentz covariance and current conservation not all of the 256 components of the tensor are independent. One finds that the tensor can be expanded using a basis of 10 Lorentz covariants which can be constructed from the vectors Q_1, Q_2 and the metric tensor $g_{\mu\nu}$. The coefficients of these covariants, the structure functions, are real functions of $Q_1{}^2$, $Q_2{}^2$ and $W_{\gamma\gamma}$. If time reversal invariance holds, 2 of these structure functions vanish. Only 6 of the remaining 8 real functions can be measured with unpolarized e^+e^- beams.

It is convenient to express the structure functions in terms of cross sections for photons with given helicities. The cross section for reaction (2.01) becomes then /12/:

$$d\sigma(e^+e^-\to e^+e^-X) = \frac{\alpha^2\sqrt{(q_1q_2)^2-q_1{}^2q_2{}^2}}{32\pi^4\ E^2\ q_1{}^2q_2{}^2} \times \frac{d^3p'_1\ d^3p'_2}{E_1{}'\cdot E_2{}'} \times \tag{2.14}$$

$$\times [\ 4\ \rho_1{}^{++}\rho_2{}^{++}\ \sigma_{TT} + 2\ \rho_1{}^{++}\rho_2{}^{00}\ \sigma_{TL} + 2\ \rho_1{}^{00}\rho_2{}^{++}\ \sigma_{LT} + \rho_1{}^{00}\rho_2{}^{00}\ \sigma_{LL}$$

$$+ 2\ |\rho_1{}^{+-}\rho_2{}^{+-}|\ \tau_{TT}\ \cos2\tilde{\varphi} - 8\ |\rho_1{}^{+0}\rho_2{}^{+0}|\ \tau_{TL}\ \cos2\tilde{\varphi} + A\cdot\tau_{TT}{}^a + B\cdot\tau_{TL}{}^a\ ].$$

The expressions for the elements of the 3×3 density matrix of the virtual photons in the $\gamma\gamma$ helicity basis, $\rho_{1,2}{}^{ab}$, are given in equation (5.13) of ref. /12/. The transformation between the two different bases (linear polarization basis and helicity basis) is derived in appendix D of the same reference. The σ's are the cross sections for unpolarized transverse photons (T) with helicities ±1 or for longitudinal photons (L) with helicity 0. Symmetry between both photons requires

$$\sigma_{TL}(W,q_1{}^2,q_2{}^2) = \sigma_{LT}(W,q_2{}^2,q_1{}^2) \tag{2.15}$$

reducing effectively the number of independent functions to be determined. Introducing in addition the cross sections $\sigma_\parallel$ and $\sigma_\perp$ for two transverse photons with linear polarization parallel or perpendicular to each other one gets the relations:

$$\sigma_{TT} = \tfrac{1}{2} \cdot (\sigma_\parallel + \sigma_\perp);$$
$$\tau_{TT} = \sigma_\parallel - \sigma_\perp. \tag{2.16}$$

The term τ_{TL} is a correlation between transverse and longitudinal photons. The coefficients of τ_{TT} and τ_{TL} both depend on the angle $\tilde{\varphi}$ between the lepton scattering planes in the $\gamma\gamma$ center of mass system. After integration over $\tilde{\varphi}$ these terms vanish. The terms $\tau_{TT}{}^a$ and $\tau_{TL}{}^a$ can only be measured with polarized lepton beams, otherwise A=B=0. All terms with an index L vanish if the corresponding photon is on-shell. Only σ_{TT} and τ_{TT} survive as both photons become real, approaching the corresponding cross sections for real photons. We obtain the following behaviour of the cross section terms as Q_1^2 and Q_2^2 approach zero:

$$
\begin{aligned}
\sigma_{TT}(W,Q_1{}^2,Q_2{}^2) \quad &\rightarrow \quad \sigma_{\gamma\gamma}{}^{unpol}(W,0,0) = \tfrac{1}{2} \cdot (\sigma_\parallel + \sigma_\perp) \\
\tau_{TT}(W,Q_1{}^2,Q_2{}^2) \quad &\rightarrow \quad \tau_{\gamma\gamma}(W,0,0) = \sigma_\parallel - \sigma_\perp \\
\sigma_{TL} \quad &\propto \quad Q_2{}^2 \\
\sigma_{LT} \quad &\propto \quad Q_1{}^2 \\
\sigma_{LL} \quad &\propto \quad Q_1{}^2 \cdot Q_2{}^2 \\
\tau_{TL} \quad &\propto \quad \sqrt{Q_1{}^2 \cdot Q_2{}^2}.
\end{aligned}
\tag{2.17}
$$

A systematic study of the amplitudes contributing in (2.14) to two-photon production of hadrons has been carried out in /14/. In principle, all cross section and correlation terms in (2.14) can be experimentally determined by varying the lepton kinematics in such a way that for fixed $W_{\gamma\gamma}$, Q_1^2 and Q_2^2, different degrees of polarization are obtained. Such an ambitious program is difficult to be realized in present experiments. First, because the variation of the photon polarization requires measurements in regions where the photon fluxes are small; secondly there are substantial problems connected with the finite resolutions of the systems which measure the lepton kinematics (tagging devices).

2.3 Approximations for the Cross Section Formula

The complex helicity structure of the cross section in (2.14) can in many cases be simplified. Because of the photon propagators in diagram a) in Fig.2.1 the radiated photons are predominantly nearly real, so that one can often make the approximation that only transverse photons contribute. The cross section (2.14) contains then only the terms σ_{TT} and τ_{TT}.

The latter term also drops out if one integrates over the relative azimuthal angles of the scattered leptons. Therefore, for a wide range of applications, the cross section for $e^+e^- \to e^+e^-X$ can be approximated by a product of the densities of transverse photons and a cross section for $\gamma\gamma \to X$:

$$d\sigma(e^+e^- \to e^+e^-X) = \frac{\alpha^2 \cdot \sqrt{(q_1 q_2)^2 - q_1^2 q_2^2}}{32\pi^4 \ E^2 \ q_1^2 q_2^2} \ 4 \ \rho_1^{++} \ \rho_2^{++} \ \sigma_{TT}(W_{\gamma\gamma}, q_1^2, q_2^2) \ \frac{d^3p'_1 \ d^3p'_2}{E_1' \cdot E_2'}.$$

(2.18)

Introducing a "two-photon luminosity function" for transverse photons, $L_{\gamma\gamma}^{TT}$, this can be rewritten as:

$$\frac{d^5\sigma(e^+e^- \to e^+e^-X)}{d\omega_1 \ d\omega_2 \ d\cos\Theta_1 \ d\cos\Theta_2 \ d\Phi} = \frac{d^5 L_{\gamma\gamma}^{TT}}{d\omega_1 \ d\omega_2 \ d\cos\Theta_1 \ d\cos\Theta_2 \ d\Phi} \cdot \sigma_{TT}(W_{\gamma\gamma}, q_1^2, q_2^2).$$

(2.19)

The differential luminosity function is:

$$\frac{d^5 L_{\gamma\gamma}^{TT}}{d\omega_1 \ d\omega_2 \ d\cos\Theta_1 \ d\cos\Theta_2 \ d\Phi} = \frac{\alpha^2}{16\pi^3} \ \frac{E_1' E_2'}{q_1^2 q_2^2} \ \sqrt{X} \ 4 \ \rho_1^{++} \ \rho_2^{++}$$

(2.20)

with the density matrix elements /12/:

$$2 \ \rho_1^{++} = X^{-1}(2p_1 q_2 - q_1 q_2)^2 + 1 + 4m_e^2/q_1^2$$
$$= X^{-1}(k - 4EE_{\gamma 2} q_2^2)^2 + 1 + 4m_e^2/q_1^2$$

(2.21)

$$\rho_2^{++} = \rho_1^{++}(1 \to 2),$$

where $k = \frac{1}{2}(W_{\gamma\gamma}^2 - q_1^2 - q_2^2)$ and $X = k^2 - q_1^2 q_2^2$.

The transverse-transverse photon luminosity function has been numerically evaluated and discussed in /15/[2]. In Fig.2.3 $dL_{\gamma\gamma}^{TT}/dz$ is plotted versus $z = W_{\gamma\gamma}/2E$ for different beam energies. $dL_{\gamma\gamma}^{TT}/dz$ is obtained by integrating (2.20) over all values of the variables which lead to the same $W_{\gamma\gamma}$. The curves A to C are calculated at different beam energies without any restriction on the scattering angles of the electrons, whereas for D and E the scattering angles of one or both electrons are restricted ("tagging", see Sect.2.4). In most analyses the fully differential form of the $\gamma\gamma$ luminosity function is used in Monte Carlo programs thus allowing the simulation of the detector efficiencies with the proper kinematics for each event.

---

[2]   Note that the (dimensionless) quantity $L_{\gamma\gamma}$ is not a luminosity proper. Defining $L' = L_{\gamma\gamma} \cdot L_{ee}$, where L_{ee} is the e^+e^- luminosity, the conventional relation between counting rate $\dot{N}$ and cross section $\sigma_{\gamma\gamma}$ is given by: $\dot{N} = L' \cdot \sigma_{\gamma\gamma}$.

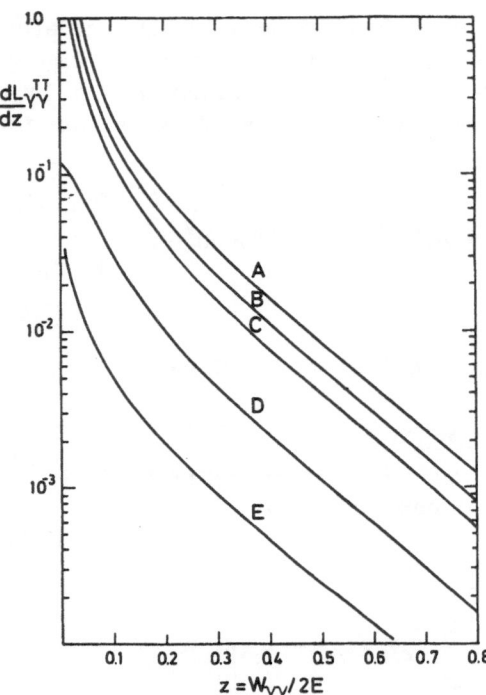

$\frac{dL_{\gamma\gamma}^{TT}}{dz}$

z = $W_{\gamma\gamma}$/2E

Fig.2.3. Differential two-photon luminosity for transverse photons /15/: Curves A, B, C give the total luminosity for E = 1000, 100, 15 GeV; curves D and E give the single and double tagged luminosity for E = 15 GeV (tagging angles: $0 \leq \Phi \leq 2\pi$, $20 < \Theta < 200$ mrad).

The functions ρ_i^{++} in (2.20) contain in general the variables of both photons. For $Q_i^2 \to 0$, $Q_i^2 \ll W_{\gamma\gamma}^2$, however, the photon luminosity function can be written as a product of two fluxes where each flux is determined by the kinematics of one vertex only. Since $Q_i^2 \ll W_{\gamma\gamma}^2$ implies $\Theta_i \ll 1$, $W_{\gamma\gamma}$ depends only on the energies of the photons: $W_{\gamma\gamma}^2 \approx 4\,E_{\gamma 1}E_{\gamma 2}$. After an integration over the angular distribution of the leptons one obtains the factorized luminosity function, which is differential in ω_1 and ω_2 ($\omega_i = E_{\gamma i}/E$):

$$\frac{d^2L_{\gamma\gamma}}{d\omega_1 d\omega_2} = \frac{dN_\gamma(\omega_1)}{d\omega_1}\frac{dN_\gamma(\omega_2)}{d\omega_2} \tag{2.22}$$

The photon spectra integrated between Q^2_{min} and $Q^2_{max} \ll W_{\gamma\gamma}^2$ take the form /12/:

$$\frac{dN_\gamma(\omega)}{d\omega} = \frac{\alpha}{2\pi}\frac{1}{\omega}\left\{[1+(1-\omega)^2]\cdot\ln\left(\frac{Q^2_{max}}{Q^2_{min}}\right) - (1-\omega)\left(1 - \frac{Q^2_{min}}{Q^2_{max}}\right)\right\}. \tag{2.23}$$

For the case that the scattered electron is detected in an angular range $m_e/E \ll \Theta_{min} < \Theta < \Theta_{max} \ll 1$ we have:

$$Q^2_{max}/Q^2_{min} = (\Theta_{max}/\Theta_{min})^2. \tag{2.24}$$

If the scattered electron is not detected Q^2_{min} reaches the kinematical limit for $\Theta = 0$ as given in (2.05):

$$Q^2_{min} = m_e^2 \cdot \omega/(1-\omega) + O(m_e^4).$$ (2.25)

Q^2_{max} is either given by the largest allowed scattering angle Θ_{max},

$$Q^2_{max} = 4EE'\sin^2(\Theta_{max}/2),$$ (2.26)

or by an effective cut-off, which for hadron production may be of the order of the ρ mass /12/:

$$Q^2_{max} \approx m_\rho^2.$$ (2.27)

Retaining only the leading term in (2.23) the photon spectrum is approximated by:

$$dN_\gamma(\omega)/d\omega = (\alpha/\pi) \cdot (1/\omega) \cdot \ln\eta \cdot [1+(1-\omega)^2]$$ (2.28)

with $\eta = E/m_e$ or $\eta = \Theta_{max}/\Theta_{min}$ depending on whether the scattered lepton is not detected at all or is detected in the range $\Theta_{min} < \Theta < \Theta_{max}$.

The differential luminosity $dL_{\gamma\gamma}/dz$ (with z= $W_{\gamma\gamma}/2E$ as defined above) is obtained by integrating (2.22) with the constraint $\omega_1 \cdot \omega_2 = z^2$. Using the approximation (2.28) for the photon spectra one obtaines:

$$dL_{\gamma\gamma}/dz = (2\alpha/\pi)^2 \cdot z^{-1} \cdot (\ln\eta)^2 \cdot f(z)$$ (2.29)

with the "Low function" f(z) /2/:

$$f(z) = (2+z^2)^2 \ln(1/z) - (1-z^2)(3+z^2).$$ (2.30)

For not too large z (z<0.8), this formula overestimates the exact luminosity function (Fig.2.3) by about 10 to 20%, but reproduces the shape of the function quite well.

The factorization in (2.22) is the "equivalent photon" or the "Weizsäcker-Williams" approximation for the two-photon case. This approximation is convenient for analytical calculations of cross sections and has been used in theoretical works and for approximate rate estimates. The various forms and the ranges of validity of the equivalent photon approximation have been extensively discussed and compared to the exact result /12, 15, 16/. In recent experimental analyses the equivalent photon approximation has hardly been used. Computer programs are available which are able to handle the multi-dimensional integration of the exact $\gamma\gamma$-luminosity function in (2.20) by Monte Carlo methods. That has also the advantage that the exact kinematics can be generated event by event and the generated events can then be passed through a detector simulation.

2.4 Special Kinematical Situations; Tagging of the Photons

In experimental studies of two-photon interactions one can "tag" the interacting photons by detecting the scattered leptons. Three different kinematical conditions can be distinguished:

1. Both scattered leptons are detected (tagging of both photons, double-tag) determining in principle the full $\gamma\gamma$ kinematics, limited only by the resolution of the tagging devices.
2. Only one scattered lepton is detected (tagging of one photon, single-tag).
3. Neither of the leptons is detected (no-tag).

As a variation of 2) and 3) the undetected leptons can be restricted to small scattering angles $<\Theta_{max}$ by requiring that no lepton is observed at angles $>\Theta_{max}$ (sometimes referred to as 'anti-tagging').

2.4.1   Double-Tagging

In principle double-tagging is the best method to measure $\gamma\gamma$ processes since it provides the complete information on the $\gamma\gamma$ system. However, it turns out that the majority of experimental results have been obtained with the no-tag or single-tag conditions. If possible, tagging is avoided because the rates drop steeply as the lepton scatters away from the extreme forward direction. Tagging at very small angles is in most cases not feasible because of background problems (small angle Bhabha scattering, beam-gas scattering). In Fig.2.3 the $\gamma\gamma$ luminosity for a given $W_{\gamma\gamma}$ can be compared for the cases that no tag is required (C), only one photon is tagged (D) and both photons are tagged (E).

Another restriction of the double-tagging method is that in practice the complete kinematical information on the $\gamma\gamma$ system cannot be obtained. Very often the resolution of $W_{\gamma\gamma}$ is not sufficient for revealing structures in the cross section. For a given $W_{\gamma\gamma}$ the resolution becomes worse for higher beam energies because the energy loss of the scattered electrons, which essentially determines $W_{\gamma\gamma}$, is measured less accurately at higher energies. See also the discussion in Sect.7.2.1.

2.4.2   Single-Tagging

The requirement of the single-tag condition is necessary in the following cases:

1. for the identification of a $\gamma\gamma$ reaction if the background of 1γ annihilation events is not small, e.g. in measurements of inclusive processes at higher $W_{\gamma\gamma}$ (total cross section measurements, study of high p_T reactions and jets) and
2. for the determination of the Q^2 dependence of resonance couplings or of the total cross section (deep-inelastic electron-photon scattering).

In the first case, where tagging is used to identify a $\gamma\gamma$ event, one wants to keep the tagging angle and thus the Q^2 of the tagged photon small in order not to loose too much rate and to avoid complications from longitudinal photon contributions (small angle tagging, if possible around 0°; at PETRA/PEP typically in the range 20-60 mrad). For the study of the deep-inelastic structure functions of the photon one wants to extent the measurements to Q^2 values as high as possible; the practical limitation is given by the achievable rate (large angle tagging: above ~200 mrad).

Of special interest (e.g. for the measurement of photon structure functions) is the case of one photon with high Q^2 interacting with a quasi-real photon (as can be defined by anti-tagging). In this case the longitudinal component can only be neglected for the quasi-real photon ($q_2^2 \approx 0$) and instead of the cross section in (2.18) we can use /12/:

$$\frac{d^5\sigma(e^+e^- \to e^+e^-X)}{d\omega_1\, d\omega_2\, d\cos\Theta_1\, d\cos\Theta_2\, d\Phi} = \frac{d^5 L_{\gamma\gamma}^{TT}}{d\omega_1\, d\omega_2\, d\cos\Theta_1\, d\cos\Theta_2\, d\Phi}$$

$$\times \left[\ \sigma_{TT}(W_{\gamma\gamma},q_1^2) + \varepsilon\sigma_{LT}(W_{\gamma\gamma},q_1^2) + \tfrac{1}{2}\varepsilon_1\varepsilon_2\tau_{TT}(W_{\gamma\gamma},q_1^2)\cdot\cos2\tilde{\varphi}\ \right] \qquad (2.31)$$

with $\varepsilon = \rho_1^{00} / (2\cdot\rho_1^{++})$
$\varepsilon_1 = |\rho_1^{+-}| / \rho_1^{++}$
$\varepsilon_2 = |\rho_2^{+-}| / \rho_1^{++}$.

Here we have allowed for the explicit measurement of the linear polarization asymmetry τ_{TT} by detecting the angle $\tilde{\varphi}$ between the scattering planes (in the $\gamma\gamma$ center of mass system). This requires a second tag at small angles.

For small Q_2^2 the luminosity function factorizes into two terms representing the fluxes of the two photons:

$$d\sigma(e^+e^- \to e^+e^-X) = \Gamma_T\left[\sigma_{TT}(W_{\gamma\gamma},q_1^2) + \varepsilon\cdot\sigma_{LT}(W_{\gamma\gamma},q_1^2) + \tfrac{1}{2}\varepsilon_1\varepsilon_2\tau_{TT}(W_{\gamma\gamma},q_1^2)\cdot\cos2\tilde{\varphi}\right]$$

$$\times\ n_\gamma\cdot dE_1'\cdot d\Omega_1\cdot d\omega_2\cdot d\tilde{\varphi}/2\pi. \qquad (2.32)$$

In this case the flux Γ_T and the polarization ε of the virtual photons depend only on the tagged lepton variables, whereas n_γ, the "photon target" density, depends only on the kinematics of the untagged photon /17/. If Θ_2 is restricted to small angles $\Theta_2 < \Theta_{2max}$, e.g. by anti-tagging, and $Q_1^2 \gg m_e^2$ we have:

$$\Gamma_T = \alpha E_1'(2-2y+y^2)/(2\pi^2\cdot Q_1^2\cdot y) \qquad (2.33)$$

$$\varepsilon = 2(1-y)/(2-2y+y^2) \qquad (2.34)$$

with $y = q_1q_2/p_1q_2 = 1-(E_1'/E)\cdot\cos^2(\Theta_1/2)$

$$n_\gamma(\omega_2,\Theta_{2max}) =$$

$$(\alpha/\pi)(1/\omega_2)\left\{[1+(1-\omega_2)^2]\cdot\ln[(E/m_e)\cdot((1-\omega_2)/\omega_2)\cdot\Theta_{2max}] -1+\omega_2\right\}. \qquad (2.35)$$

The form of the cross section in (2.32) is analogous to the electron-nucleon scattering cross section. Correspondingly one can define an electron-photon scattering cross section by:

$$\frac{d\sigma(e\gamma \to eX)}{d\Omega_1 \cdot dE_1' \cdot d\tilde{\phi}/2\pi} = \Gamma_T \left[\sigma_{TT}(W_{\gamma\gamma}, q_1^2) + \varepsilon \cdot \sigma_{LT}(W_{\gamma\gamma}, q_1^2) + \tfrac{1}{2}\varepsilon_1\varepsilon_2 \tau_{TT}(W_{\gamma\gamma}, q_1^2) \cdot \cos 2\tilde{\phi} \right].$$

$$(2.36)$$

For large Q_1^2 we talk about deep-inelastic electron-photon scattering. As in deep-inelastic lepton-nucleon scattering the cross section can be expressed in terms of structure functions. These structure functions will be introduced in Chap.9.

2.4.3   No-tag Experiments

Experimental experience has shown that tagging is usually not necessary for studying exclusive final states produced by quasi-real photons. For the rejection of $\gamma\gamma$ final states with undetected particles and other backgrounds, one takes advantage of the following feature of $\gamma\gamma$ reactions: Because the photons are dominantly radiated along the beam direction, the transverse momentum of the $\gamma\gamma$ system with respect to the beams is preferentially small. Reconstructing the total transverse momentum from

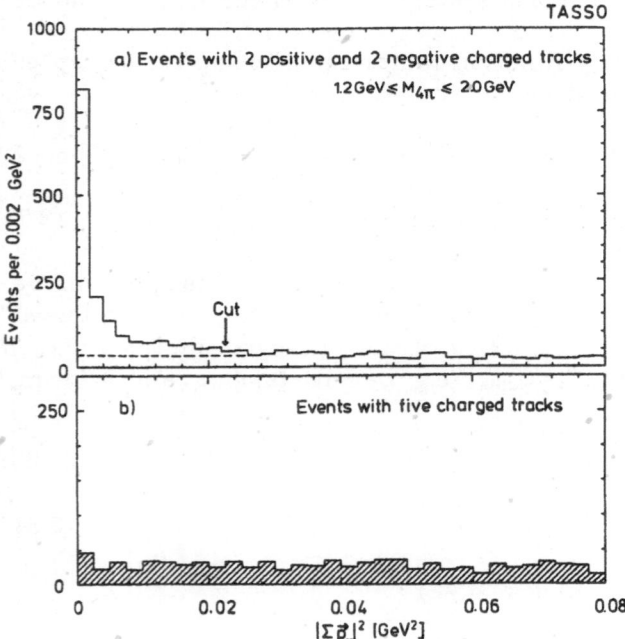

Fig.2.4.   Distribution of the square of the transverse momentum of the detected particle system : a) $|\sum \vec{p}_T|^2$ for events with four particles detected,   b) $|\sum \vec{p}_T|^2$ for events with five particles detected where at least one charged particle is missing.

the detected particles, $\left|\sum\vec{p}_T\right|$, one finds that the $\left|\sum\vec{p}_T\right|^2$ distribution has a sharp peak at small $\left|\sum\vec{p}_T\right|^2$ if all particles of the final state are measured, whereas the same distribution is flat near 0 if particles are missing. This is demonstrated in Fig.2.4 where $\left|\sum\vec{p}_T\right|^2$ is plotted a) for measured events with four charged particles and total charge 0 and in b) for events with five charged particles detected. A cut in $\left|\sum\vec{p}_T\right|$, typically at about 100 MeV, preferentially selects the completely reconstructed events and effectively restricts the Q^2's of the photons to small values. Neglecting the rare cases where both photons have large opposite transverse momenta, the largest Q^2's are roughly of the order of the maximum $\left|\sum\vec{p}_T\right|^2$.

In the analyses of $\gamma\gamma$ exclusive final states in the resonance region completed so far at PETRA/PEP energies, the background contributions from 1γ annihilation reactions turned out to be negligible.

3. Detectors for Two-Photon Physics

This chapter summarizes some information on the detectors mentioned later in the discussions of experimental results. Those features of the detectors which have been relevant for the analysis of two-photon interactions will be emphasized. We start with some general remarks on the detection of two-photon reactions.

3.1 Detectors at e^+e^- Storage Rings

At e^+e^- storage rings the detectors are usually designed to meet in particular the needs of one-photon physics. Most of the detectors have a magnetic field parallel to the beam with cylindrical wire chambers and time-of-flight (TOF) scintillation counters within the field volume (see Fig.3.1). The chambers typically cover 80 to 90% of the total solid angle leaving acceptance holes around the beam line. The energies of photons and electrons are measured in electromagnetic shower calorimeters covering as large a solid angle as possible. If the shower detectors are behind the coil of the magnet, which is usually the case, the efficiency for photon

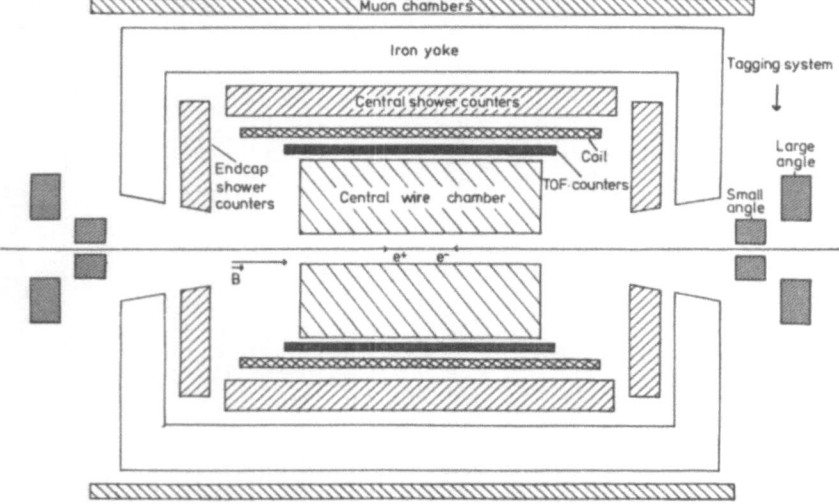

Fig.3.1.   Typical e^+e^- storage ring detector.

detection is small below about 100 MeV photon energy, making reconstruction of the typically low-energy $\gamma\gamma$ events a difficult job (see Fig.3.9).
A detector which is in some sense orthogonal to these typical storage ring
detectors is the Crystal Ball detector which emphasizes good detection of
electromagnetic showers (see description below).

The hardware triggers selecting events from e^+e^- reactions require
charged tracks in the wire chambers or neutral energy in the calorimeters
or both. The charged trigger particles have to have some minimum transverse momentum with respect to the beam, typically between 100 and 300
MeV. A trigger based on the total neutral energy deposited in the shower
counters has to require at least 2 to 3 GeV in the PETRA/PEP environment.
The threshold can be reduced by demanding separated clusters of energy,
which in addition may have to fulfil some topology requirement. The JADE
group, for example, showed that even at PETRA energies it is possible to
trigger on $\gamma\gamma$ events with only photons in the final state, like
$\gamma\gamma \rightarrow f(1270) \rightarrow \pi^0\pi^0 \rightarrow \gamma\gamma\gamma\gamma$.

Two characteristics of two-photon reactions make it particularly difficult to construct a detector well suited for both one-photon and
two-photon physics:

1. Most of the two-photon reactions occur at <u>low energy</u>. The higher the
    beam energy the larger is the disparity between the requirements for
    triggering on low energy two-photon events and on high energy
    one-photon annihilation events.
2. Two-photon events have in general a <u>boost</u> in the direction of one of
    the incoming beams, because the photon energies can be unequal. In
    contrast, for one-photon annihilation events, the center of mass sys-
    tem and the laboratory system coincide.

The low energy of $\gamma\gamma$ events is advantageous when it comes to separate
one-photon and two-photon events (see Fig.3.2), but it has a negative impact on the trigger and detection efficiencies, especially if a large

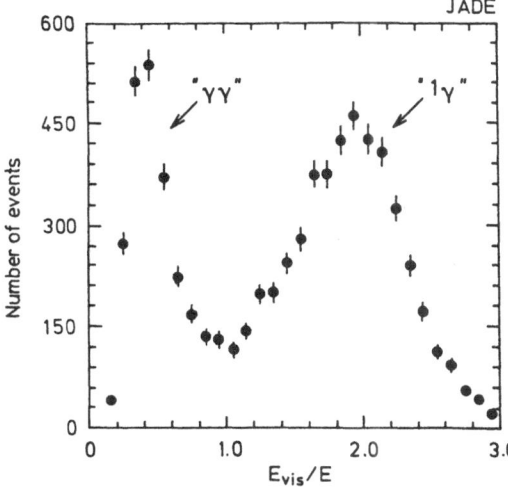

Fig.3.2. Distribution of the observed energies, E_{vis}, in e^+e^-
interactions (normalized to the
beam energy) as measured with
the JADE detector. (The $\gamma\gamma$ contribution is relatively suppressed
by the requirement that the observed particle system has a
small boost in beam direction).

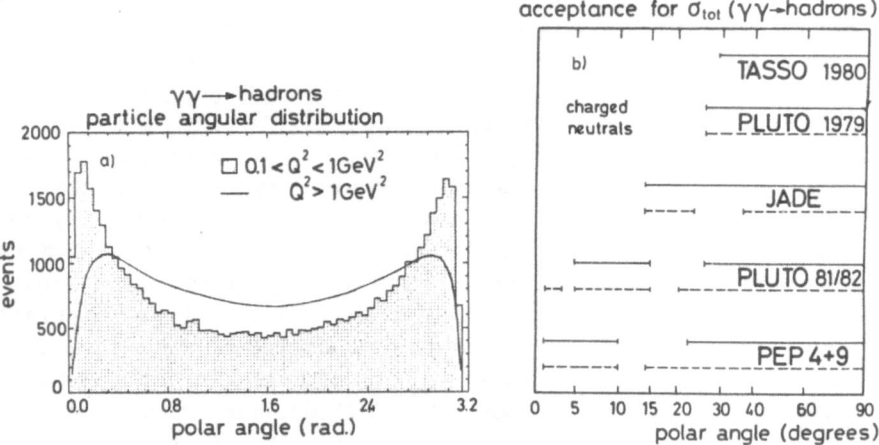

Fig.3.3. Angular distributions and detector acceptances for $\gamma\gamma$ events: a) Single hadron polar angular distribution in the laboratory system as obtained from a Monte Carlo simulation of $\gamma\gamma\to$hadrons (see Chap.7). At low Q^2 a VMD model (histogram) and at high Q^2 the Born approximation for quark pair production (curve) is used. b) Polar angular acceptance of some detectors at PETRA and PEP for charged (full lines) and neutral particles (dashed lines). The abscissa has been transformed so that a constant particle flux per unit length of the abscissa is obtained for the VMD model used in a).

low-energy background (like beam-gas scattering) requires the trigger thresholds to be raised.

The boost of the event in the forward direction leads to high particle densities in the region close to the beam where all detectors have acceptance holes. The situation is especially bad in the low-Q^2 region where the cross section is dominated by peripheral scattering according to the VMD picture. In such a reaction the particles are produced in the forward direction with limited transverse momenta with respect to the beam. Fig.3.3a shows the particle distribution as a function of the angle Θ between the particle and the beam. The hatched histogram is the expectation from the VMD model for low Q^2, while the curve is calculated for large Q^2 using the Born approximation for quark pair production with subsequent fragmentation of the quarks into hadrons. The Θ ranges accepted by the different detectors are plotted in Fig.3.3b. In order to demonstrate the importance of the forward region the Θ scale is transformed so that a constant length on the Θ axis corresponds to a constant particle flux (in the VMD model). An optimal Θ coverage is obtained by the two detectors specialized for two-photon physics, the PEP-9 detector at PEP (Fig.3.4, /19/) and the PLUTO detector as operated in 1981-82 (Fig.3.5). These detectors have special forward spectrometers with additional magnetic fields perpendicular to the beam, which allow the measurement of charged particle momenta in the forward direction.

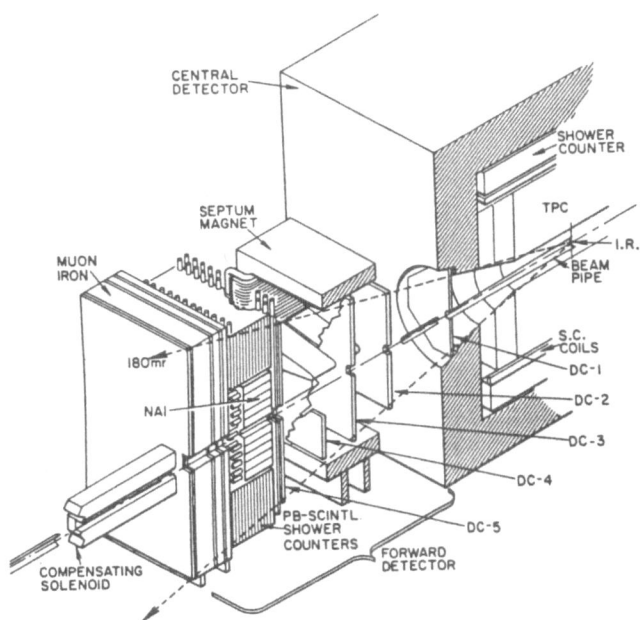

<u>Fig.3.4.</u>   The specialized two-photon detector PEP-9: The detector consists of two forward spectrometers, one on each side of the inter- action region (I.R.). The central detector is the time projection cham- ber (TPC) of the PEP-4 experiment surrounded by a superconducting (S.C.) coil of a solenoidal magnet. DC and NAI denote drift chambers and NaI shower counters, respectively.

3.2 Tagging Devices

Some goals of two-photon physics require the detection of one or both scattered electrons (tagging, see Sect.2.4). Reasons for using the tagging method can be:

- to achieve a complete kinematical reconstruction of an event (e.g. for total cross section measurements),
- to suppress background,
- to determine the Q^2 value of the tagged photon.

Depending on the purpose, the tagging detectors have to cover different angular ranges in the forward direction, see Fig.3.1. Tagging of quasi-real photons can be done without a large reduction of the counting rate by de- tecting the scattered electrons at very small angles close to 0°. 0° tagging has been realized at the storage ring ADONE /6/, at DORIS with the BONAN- ZA experiment /20/ and at DCI with the DM1 detector /21/. The detection of electrons scattered around 0° is only possible if they are deflected from the beam. That can be done employing the magnets of the beam transport system near the interaction region. Placing appropriate detectors for the

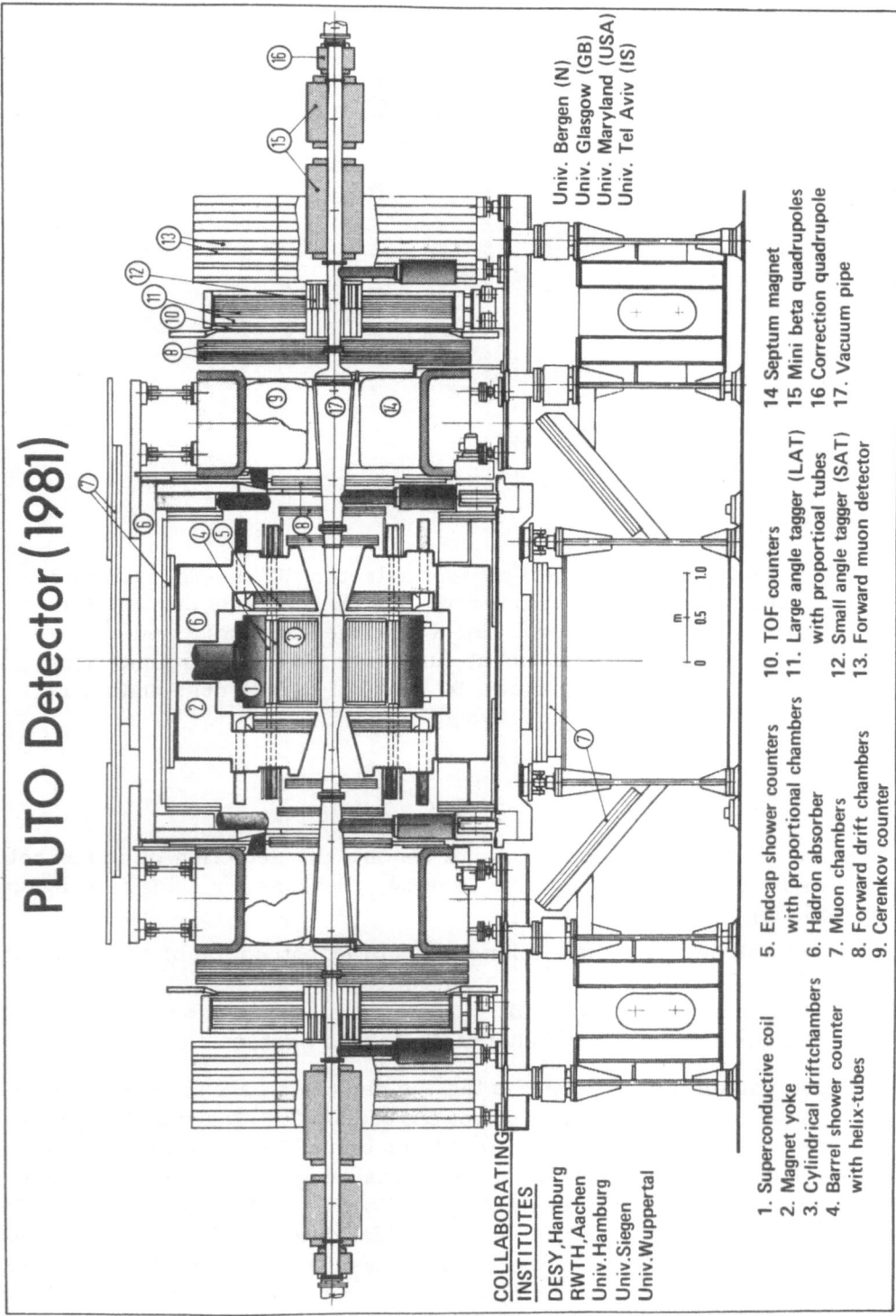

PLUTO Detector (1981)

COLLABORATING
INSTITUTES

DESY, Hamburg
RWTH, Aachen
Univ. Hamburg
Univ. Siegen
Univ. Wuppertal

Univ. Bergen (N)
Univ. Glasgow (GB)
Univ. Maryland (USA)
Univ. Tel Aviv (IS)

1. Superconductive coil
2. Magnet yoke
3. Cylindrical driftchambers
4. Barrel shower counter
   with helix-tubes
5. Endcap shower counters
   with proportional chambers
6. Hadron absorber
7. Muon chambers
8. Forward drift chambers
9. Cerenkov counter

10. TOF counters
11. Large angle tagger (LAT)
    with proportioal tubes
12. Small angle tagger (SAT)
13. Forward muon detector

14. Septum magnet
15. Mini beta quadrupoles
16. Correction quadrupole
17. Vacuum pipe

m
0   0.5   1.0

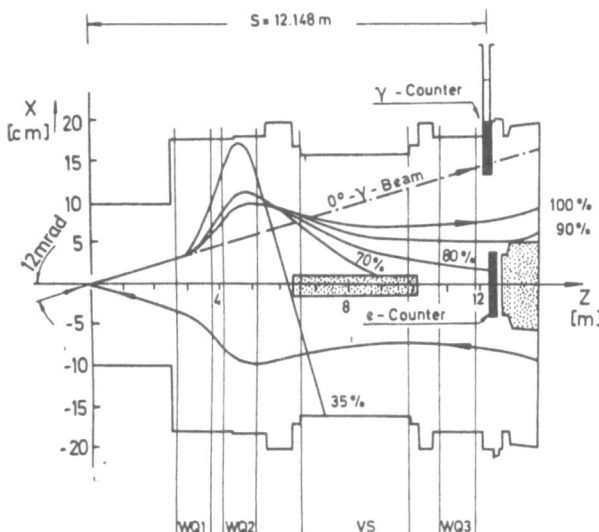

<u>Fig.3.6.</u>   Tagging system used by the BONANZA experiment at DORIS: Shown is a plane vertical to the storage ring containing the the beams which intersect at (x=0, z=0). The curve in the lower half and the curve labelled "100%" in the upper half present the incoming and (non-interacting) outgoing beams, respectively. The other curves are the trajectories for beam particles coming from the interaction point with energies reduced to certain fractions of the beam energy (90%, 80%, 70%). WQ1 to WQ3 denote the positions of quadrupoles and VS of a septum magnet. The dotted areas indicate absorbers.

electrons outside the beam line, this system can be used as a spectrometer. As an example we show the 0° tagging facility of the BONAN-ZA experiment at DORIS /20/, which used advantageously the double-ring structure of this machine (DORIS had two rings until 1980). Because the beams have a vertical crossing angle at the interaction region, they pass off center through the quadrupoles WQ1 and WQ2 resulting in a large ener-gy dispersion (Fig.3.6). The scattered electrons are detected in shower counters mounted inside the beam pipe 12 meters away from the inter-action point (e-counter). The counters accept electrons with energies between 73% and 82% of the beam energy and with scattering angles up to 10 mrad. Similar counters (γ-counter) are used to veto beam-gas bremstrahlung events.

    The 0° tagging system of the DCI storage ring is described below togeth-er with the DM1 detector (Fig.3.8). Recently, the ARGUS group proposed to install a 0° tagger into their experiment at DORIS II /22/. DORIS II has only

---

<u>Fig.3.5.</u>   The PLUTO detector as operated in 1981-82 as a dedicated detector for two-photon physics.

one ring, but the vertical bending near the interaction point is the same as in the old double ring structure. Using BGO crystals (BGO = bismuth germanium oxide) for the measurement of the electron energy, the resolution is expected to be sufficient to measure the total cross section in a double-tag experiment even in the resonance region.

The dominant background in 0° tagging devices comes from electrons which loose energy by beam-gas bremsstrahlung. Small angle Bhabha scattering can be kinematically suppressed. At PETRA/PEP energies the background is probably too high to apply 0° tagging (see e.g. /23/). At these machines, tagging starts typically at 20 mrad, where a deflection of the scattered electrons away from the beam is no longer necessary. The tagging devices are usually shower calorimeters with resolutions σ_E/E between about $3\%/E^{1/4}$ for NaI (e.g. PEP-9 detector) and about $17\%/\sqrt{E}$ (E in GeV) for lead- scintillator sandwiches with wavelength shifter readout.

The study of the Q^2 dependence of cross sections requires tagging over a large range. Some detectors have a special large-angle tagger covering the range from about 100 to 200 mrad (Q^2 up to 9 GeV² for a beam energy of 15 GeV), see e.g. the PLUTO detector in Fig.3.5. Above ~200 mrad the endcap calorimeters and even the central detectors (starting at about 20° to 30°) are used for tagging.

3.3 Some Remarks on Particle Identification

A separation of pions from electrons and muons is desirable for suppressing background from QED reactions, especially in channels with two charged tracks in the final state, such as $\gamma\gamma \rightarrow \pi^+\pi^-$. The separation using shower counters and muon filters requires momenta larger than about 1 GeV. The Mark II group applied a technique to enhance the pion pair signal exploiting the pion absorption in the magnet coil in front of a calorimeter (see Sect.6.4.2). An interesting method, though limited to low energy machines, has been used with the DM1 detector at DCI by kinematically reconstructing the masses of the produced particles (see Sect.4.2).

Time-of-flight measurements with scintillation counters surrounding the central detector have been successfully used to identify charged kaons (up to momenta of about 1 GeV) and protons (up to about 1.5 GeV). Figure 3.7 shows in a 3-dimensional scatter plot the masses calculated from time-of-flight for the negatively and positively charged particles in events where two tracks have been detected. The large truncated peak contains electron, muon and pion pairs, the slim spike is due to kaon pairs and the cluster at high masses comes from proton-antiproton pairs. Charged particles with low momenta have also been identified employing energy loss measurements with the central chambers, such as with the JADE "jet" chamber or the TPC (=time projection chamber) at PEP.

The detection of neutral energy in shower counters covering a large solid angle is important for the reconstruction of $\pi°$'s and η's (especially for resonance physics) as well as to complete the reconstruction of the final

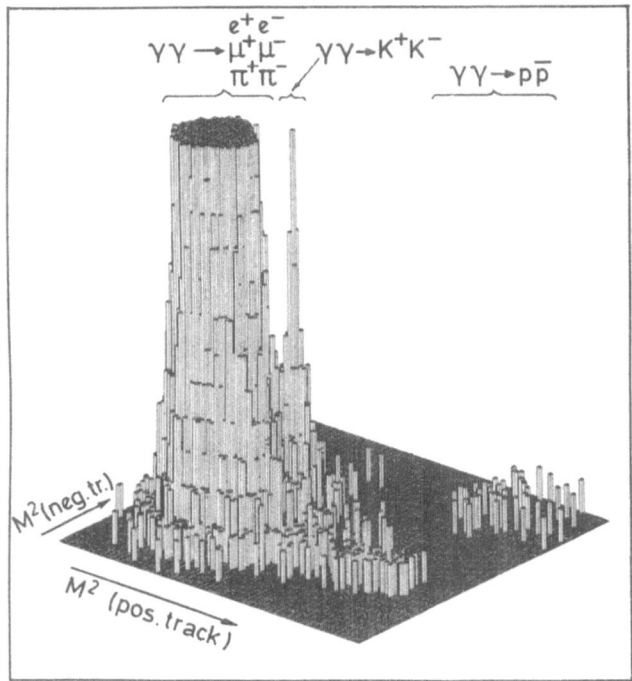

<u>Fig.3.7.</u>   Particle identification by time-of-flight measurement: Scatter plot of the square of the mass calculated from time-of-flight (M^2). M^2 is plotted for the positive and negative charged particle of events with two detected tracks with momenta smaller than 0.9 GeV (TASSO).

state. The latter is of extreme importance since the total center of mass energy of a $\gamma\gamma$ event is in general unknown.

3.4 Description of the Detectors

In the following we summarize the main properties of those detectors which have contributed in recent years to the experimental results in two-photon physics. The detector components which are important for the analyses discussed in this review are emphasized. The detectors are running or had been running at the following storage rings:

 — DCI at Orsay (DM1), beam energies up to about 1 GeV,
 — SPEAR at Stanford (Mark II, Crystal Ball, SP-14), beam energies up to about 3.7 GeV,
 — PETRA at Hamburg (CELLO, JADE, Mark J, PLUTO, TASSO), beam energies up to about 19 GeV.

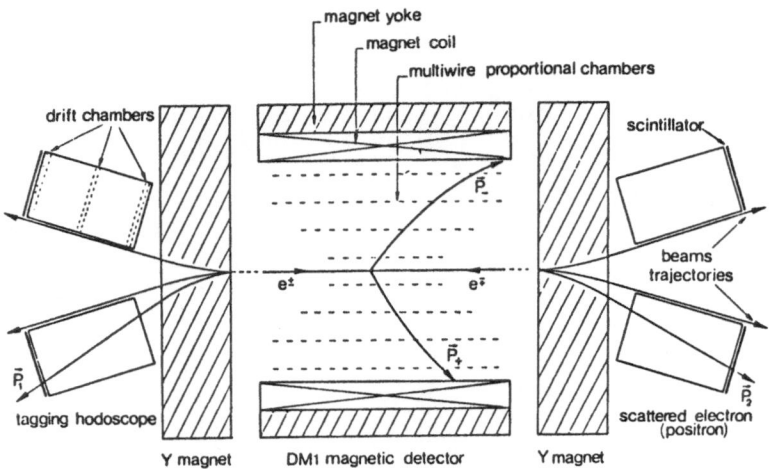

<u>Fig.3.8.</u>   DM1 detector with 0° tagging system.

The maximum energies given here correspond to the running periods in which the data presented in this article were taken. Experiments at the PEP storage ring have not been included, because only preliminary results on two-photon physics have been reported up to now.

The <u>DM1 DETECTOR</u> /21/ at the storage ring DCI is a magnetic detector (Fig.3.8). Charged particles are detected with cylindrical wire chambers in 67% of 4π sterad and for transverse momenta larger than 75 MeV. No particle identification is possible. The 0° tagging system uses the bending magnets of the DCI double ring, which are closest to the interaction region, as spectrometers. Electrons scattered at angles between 0 and 10 mrad are detected if they have an energy loss between 20% and 53%. The momentum and angular resolution of the tagging system are 1% and 1 mrad, respectively.

The data reported from the <u>MARK II DETECTOR</u> were taken at the storage ring SPEAR. The Mark II detector /24/ consists of a cylindrical drift chamber surrounded by time-of-flight scintillation counters (75% of 4π sterad coverage). Both are embedded in a solenoidal magnet with a field of 0.41 Tesla. Liquid argon shower calorimeters (65% of 4π sterad) and a muon detection system (55% of 4π sterad) are located outside the magnet coil. Shower counters also cover both ends of the detector around the beam. The resolution of the time-of-flight system is 300 psec, leading to a one standard deviation π-K and K-p separation at momenta of 1.35 GeV and 2.0 GeV, respectively. The resolution of the liquid argon calorimeter is $\sigma_E/E = 11\%/\sqrt{E}$ (E in GeV) for E>0.5 GeV, with photon detection efficiencies of 15% at 100 MeV, 50% at 200 MeV and >90% above 400 MeV for photons within the geometrical acceptance.

The results reported here from the <u>CRYSTAL BALL DETECTOR</u> have been obtained at the storage ring SPEAR. The Crystal Ball /25/ is a nonmagnetic detector consisting of a spherical array of NaI(Tl) shower counters covering 93% (98% including endcaps) of 4π sterad. Photon detection is in principle possible down to the keV region, but is in practice limited to about 10 MeV due to background noise. The energy and angular resolution for electromagnetic showers are $\sigma_E/E = 2.6\%/E^{1/4}$ (E in GeV) and $1°-2°$, respectively. The directions of charged particles are determined by a central wire chamber. Charged particle momenta are not measured and so the Crystal Ball works best for the detection of final states with photons and electrons only.

The apparatus of the SP-14 EXPERIMENT at SPEAR /26/, which was used for a double tag measurement, also has no magnetic field. It has four identical quadrants each consisting of two parts: 1) A small-angle detector (for tagging) made of NaI crystals covering an angular range from 55 to 180 mrad and having an energy resolution of 3.8% (fwhm) around 3 GeV. 2) A detector covering a polar angular range between $15°$ and $30°$ which allows a separation of electrons, muons and hadrons. A crude central detector provides information on the azimuth of charged particle tracks in a wide polar angular range ($22°$ to $158°$).

The CELLO DETECTOR at PETRA /27/ has a superconducting solenoidal magnet with a field of 1.3 Tesla. Charged tracks are measured by a cylindrical wire chamber covering 93% of 4π sterad. The central liquid argon calorimeter is placed outside the thin coil (0.48 radiation lengths) and covers 88% of the solid angle. The endcap calorimeters cover the polar angular range $0.91<|\cos\theta|<0.99$. The energy resolution of the calorimeters is $\sigma_E/E = 13\%/\sqrt{E}$ (E in GeV).

The JADE DETECTOR /28/ at PETRA is a magnetic (0.48 Tesla) detector with a central drift chamber allowing to detect charged tracks within 97% of 4π sterad. The chamber provides also a dE/dx measurement for charged particle identification. Electromagnetic showers are detected in 90% of 4π sterad by lead glass arrays surrounding the central detector outside the coil and in the endcap lead glass counters (245 to 500 mrad). Figure 3.9 shows the energy dependence of the electron and photon detection efficiencies. The whole apparatus is surrounded by a muon identification system. A small-angle tagger covers the angular range from 34 to 75 mrad.

The MARK J DETECTOR /29/ at PETRA is a calorimeter designed to measure electrons, muons, neutral particles and charged hadrons within a polar angular range from $9°$ to $171°$. A magnetized iron filter is used to identify muons and to measure their momenta.

In the following the PLUTO DETECTOR /30/ is described as it was in operation at PETRA until 1979 (Fig.3.10). The central detector consists of cylindrical proportional chambers in a magnetic field of 1.65 Tesla covering 87% of 4π sterad. The central shower counters ($\sigma_E/E = 35\%/\sqrt{E}$, E in GeV) and the endcap shower counters ($\sigma_E/E = 19\%/\sqrt{E}$) cover 94% of 4π sterad. A small-angle tagger detects electrons between 23 and 70 mrad with a lead glass shower counter matrix of 96 blocks ($\sigma_E/E = 8.5\%/\sqrt{E}$). A large-angle tagger, made of lead-scintillator sandwich counters, covers the angular region between 70 and 260 mrad ($\sigma_E/E = 11\%/\sqrt{E}$). In both taggers position

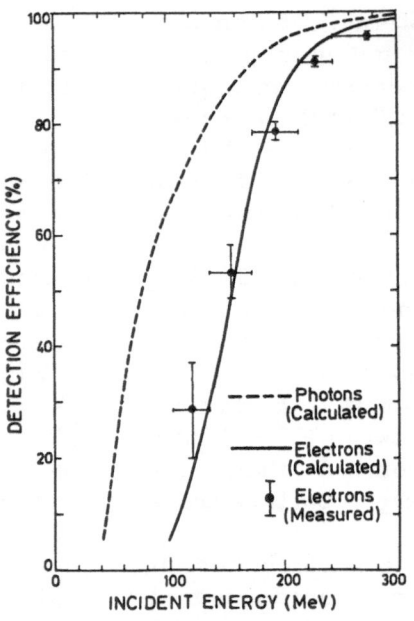

Fig.3.9.   Energy dependence of the photon and electron detection efficiencies (JADE).

measurements are provided by wire chambers in front of the shower counters.

In 1981-82 the PLUTO detector was set up for a dedicated two-photon experiment (Fig.3.5). For this purpose, forward spectrometers with magnetic dipole fields perpendicular to the beams have been added on both sides. The spectrometers are equipped with drift chambers, shower counters and muon detectors. Charged particle momenta can be measured with a resolution $\sigma_p/p = 0.025 \cdot p$ (p in GeV) in a polar angular range from 5° to 15°.

The TASSO DETECTOR /31/ is a magnetic detector (0.5 Tesla) with central tracking chambers covering 87% of 4π sterad. The 48 scintillation counters surrounding the beam at a radius of 132 cm allow particle iden-

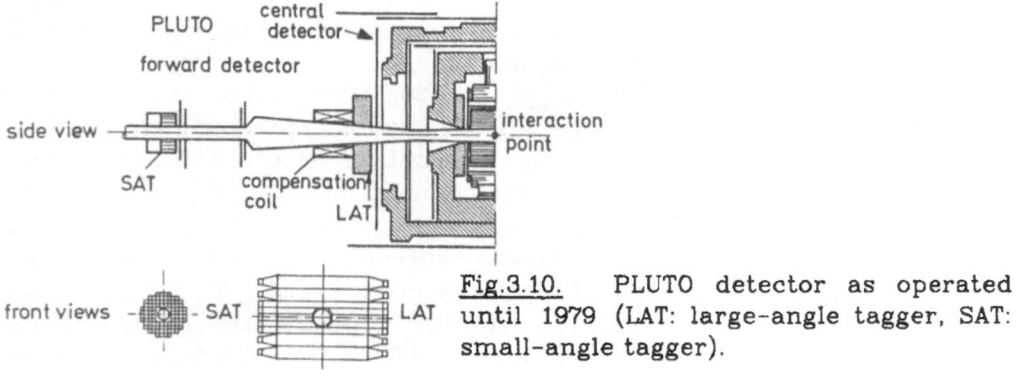

Fig.3.10.   PLUTO detector as operated until 1979 (LAT: large-angle tagger, SAT: small-angle tagger).

tification by time-of-flight measurement within 82% of 4π sterad. The r.m.s. time resolution of the counters is on the average 380 psec. Showering particles are detected in the central liquid argon shower calorimeter (44% of 4π sterad), in liquid argon endcaps (polar angular range from 11° to 28°) and in lead–scintillator shower counters in two arms at large polar angles (20% of 4π sterad). For the analyses of the $\gamma\gamma$ total cross section (Chap.7) and of $\gamma\gamma$ jets (Chap.8) forward detectors made of lead glass blocks covering an angular range between 24 and 60 mrad were used for tagging.

4. Two-Photon QED Processes

4.1 QED Cross Sections

The classical linear equations of electromagnetism do not allow scattering of electromagnetic waves, because the linear superposition principle requires that the "rays transverse one another without hindrance" (Christiaan Huygens, /32/). The quantized theory of electromagnetism (QED), however, predicts nonlinear interactions of electromagnetic waves arising from the creation and absorption of virtual fermion pairs /1/. Therefore, elastic photon-photon scattering is possible via the fermion box in the diagram of Fig.4.1. Including only the electron box, the cross section below the threshold for e^+e^- pair creation is given by /1, 33/:

$$\sigma(\gamma\gamma\rightarrow\gamma\gamma) = 129\cdot(W_{\gamma\gamma}{}^2/4m_e{}^2)^3 \text{ nb}. \tag{4.01}$$

For visible light the cross section is about 10^{-29} nb. The measurement of such a small cross section looks hopeless since even the most powerful lasers would only provide a luminosity of the order of 10^{10} to 10^{12} $\text{nb}^{-1}\text{sec}^{-1}$. Nevertheless, long before lasers became available, attempts had been made to measure light-light scattering. With the "dark-adapted eye" as detector for the scattered light, an upper limit for the cross section of the photon of $3\cdot10^{-20}$ cm² has been obtained /34/. At much higher photon energies, elastic $\gamma\gamma$ scattering has meanwhile been observed in the resonant channel (/35/, see Chap.6):

$$\gamma\gamma \rightarrow \eta \rightarrow \gamma\gamma. \tag{4.02}$$

If the two-photon invariant mass exceeds twice the mass of a lepton both elastic scattering and production of real lepton pairs is possible. Lepton pair production has been observed in the reaction:

$$e^+e^-\rightarrow e^+e^- + l^+l^- \quad (l = e, \mu). \tag{4.03}$$

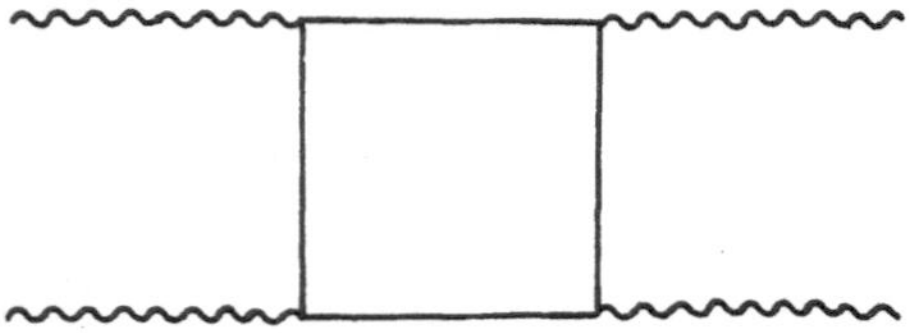

Fig.4.1. Diagram for elastic $\gamma\gamma$ scattering via a fermion loop.

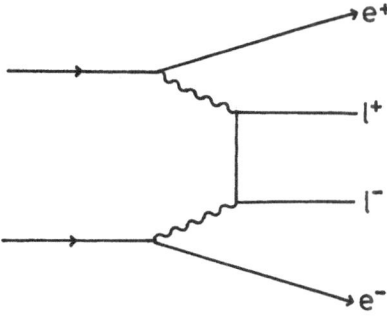

<u>Fig.4.2.</u>   Diagram for lepton pair production by two photons.

This reaction is dominated by the graph in Fig.4.2 which is of order α^4. The cross sections and angular distributions of reaction (4.03) have been calculated by many authors /e.g. 12, 36/ combining the equivalent photon approximation (Sect.2.3) with the two-photon cross section. The total and differential cross sections for lepton pair production by two real photons are given by:

$$\sigma(\gamma\gamma\to l^+l^-) = 2\pi\cdot\alpha^2\cdot W_{\gamma\gamma}^{-2}[2\beta^{*3}-4\beta^*+(3-\beta^{*4})\cdot\ln\{(1+\beta^*)/(1-\beta^*)\}]$$

$$\tag{4.04}$$

$$d\sigma/d\cos\vartheta^* = 2\pi\cdot\alpha^2\cdot W_{\gamma\gamma}^{-2}\beta^*[2\beta^{*2}\sin^2\vartheta^*-\beta^{*4}\sin^4\vartheta^*+1-\beta^{*4}]/(1-\beta^{*2}\cos^2\vartheta^*)^2.$$

The polar angle ϑ^* with respect to the $\gamma\gamma$ direction and the velocity β^* of a lepton are both defined in the $\gamma\gamma$ center of mass system. The total cross section for reaction (4.03) is obtained by convoluting (4.04) with the two-photon luminosity (2.29) and is approximately /37/:

$$\sigma(e^+e^-\to e^+e^-l^+l^-) = (28\alpha^4/27\pi m_l^2)\{\ln(4E^2/m_e^2)\}^2\cdot\ln(4E^2/m_l^2). \tag{4.05}$$

In Fig.4.3 cross sections for e^+e^- scattering are plotted. Lepton pair production via the two-photon process is compared to $\mu^+\mu^-$ production via one-photon annihilation. The curve for $\pi^+\pi^-$ production represents the QED calculation for pointlike scalar pions. Though the total two-photon cross section for e^+e^- production is orders of magnitude larger than the $\mu^+\mu^-$ cross section, they become almost equal in the kinematical range accessible to the experiments. The e^+e^- cross section has large contributions from small invariant masses and small angles due to the lepton propagator in the diagram of Fig.4.2. Therefore the experimental situation is better taken into account if one compares the differential cross sections at large angles and above some threshold for the invariant mass of the produced lepton pairs as in Fig.4.4. This figure shows also the one-photon $\mu^+\mu^-$ production cross section, demonstrating the different dependence of the two processes on the beam energy.

For a comparison of experimental results with the exact $O(\alpha^4)$ QED predictions (in contrast to the equivalent photon approximation), a Monte Carlo integration program has been written by Vermaseren /38/. The pro-

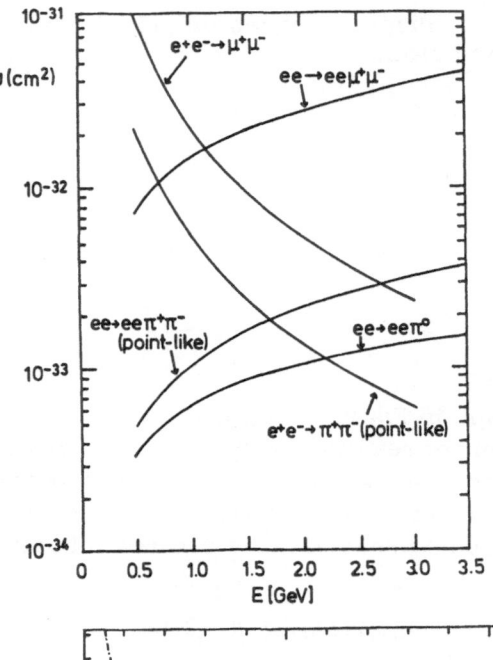

Fig.4.3.   Examples for production cross sections for e^+e^- colliding beams as a function of the beam energy (taken from /12/).

Fig.4.4.   The differential cross section $d\sigma(e^+e^-\to e^+e^-l^+l^-)/d\cos\vartheta^*$ at $\vartheta^*=90°$ (l=e or μ) for 0.5 GeV$<W_{\gamma\gamma}<2\cdot E$ plotted versus the beam energy E. The two-photon cross sections are compared to the one-photon cross section $d\sigma(e^+e^-\to\mu^+\mu^-)/d\cos\vartheta^*$ at $\vartheta^*=90°$.

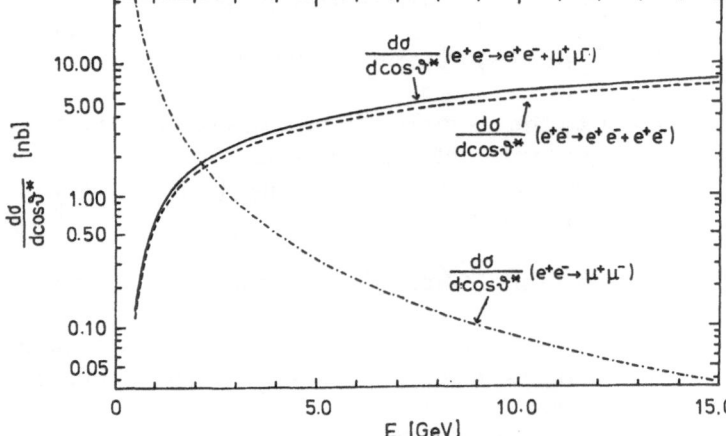

gram includes all graphs of order α^4 which lead to reaction (4.03) (Fig.4.5). The diagrams in a), which are the two-photon reactions with C-even final states, dominate in most cases. The diagrams in b), though for most experimental situations negligible, can give a sizeable contribution at small invariant lepton pair masses and/or large tagging angles. Interference between the diagrams a) and b) leads to charge asymmetries of the produced lepton pairs with respect to the incoming leptons. The annihilation diagrams in c) are suppressed because they involve only timelike photons. The Vermaseren program generates events which have to be used with weights. Though an attempt is made to keep the fluctuation of the weights small by using so-called "importance sampling" methods, certain kinematical regions (in particular for electron pair production) are

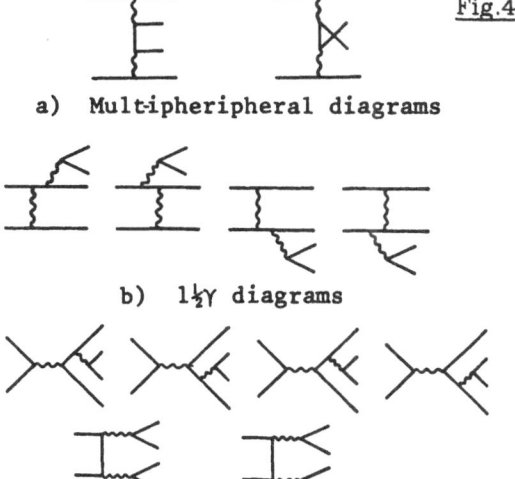

<u>Fig.4.5.</u>   Diagrams for the reaction
$e^+e^- \to e^+e^- \mu^+ \mu^-$.

a)   Multipheripheral diagrams

b) $1\frac{1}{2}\gamma$ diagrams

c)   Annihilation diagrams

problematic. Recently, a Monte Carlo program for two-photon QED processes has been presented which has the advantage of generating
unweighted events and which includes also radiative corrections /39/.

The QED reactions (4.03) were first observed at the e^+e^- storage rings in
Novosibirsk and Frascati /5, 6/. These early experiments have been reviewed in /40/. In recent years, experimental activities have concentrated
more on the hadronic final states in two-photon scattering. Nevertheless,
QED reactions are still under study, mainly for two reasons: First, in many
analyses it is necessary to understand the QED processes quite accurately
because they contribute to the background (especially in the analysis of
the $\pi^+\pi^-$ final state). Secondly, there is a special interest in the process

$$e^+e^- \to e^+e^- \mu^+ \mu^- \tag{4.06}$$

as a prototype reaction for quark pair production

$$e^+e^- \to e^+e^- q\bar{q}. \tag{4.07}$$

At high energies reaction (4.07) leads to two jets of hadrons in the final
state. In the kinematical regime where the pointlike nature of the photon
is expected to show up, i.e. at high momentum transfers, a comparison of
two jet production and muon pair production should be meaningful (see
Chap.8,9).

4.2 Measurements

A group at the storage ring DCI reported results on two-photon production of charged particle pairs (e, μ, π) in the region from threshold up
to 0.8 GeV /41/. For beam energies between 0.7 and 1 GeV they used the 0°

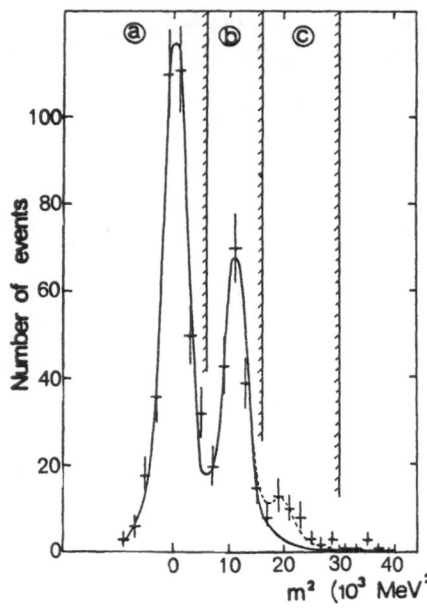

Fig.4.6.   Measurement of two-photon production of charged particle pairs (single and double tag) /41/: Distribution of the mass squared of the particles which belong to a pair (assuming that both particles of the detected pair have the same mass). The mass is determined from the kinematics of the tagged photon and of the particles measured in the central detector. The regions of electrons (a), muons (b), and pions (c) are indicated.

tagging system of the DM1 detector (see Fig.3.8). The produced particle pairs were detected in the central magnetic detector down to transverse momenta of 75 MeV. Direct particle identification was not possible. However, assuming equal masses for both particles, the mass of a produced particle could be reconstructed from kinematics for both single and double tag events. The measured mass spectrum is shown in Fig.4.6. The relative rate of accepted electron and muon pairs was measured to be

$$N_{ee}/N_{\mu\mu} = 0.524 \pm 0.049,$$

in very good agreement with the QED prediction of 0.522. The results on pion pair production from this experiment will be discussed in Sect.5.1 and Sect.6.5.

The Mark J collaboration studied the spectra of identified muon pairs with momenta larger than 1.5 GeV (for at least one μ) /42/ and compared them to the theoretical predictions for $e^+e^- \rightarrow e^+e^-\mu^+\mu^-$ as obtained from the Vermaseren program. In Fig.4.7a the observed cross section is plotted versus the e^+e^- center of mass energy. The agreement with the QED prediction is good. Due to detector efficiencies the observed cross section for $e^+e^- \rightarrow e^+e^-\mu^+\mu^-$ becomes only at the highest energies larger than the cross section for the one-photon annihilation process $e^+e^- \rightarrow \mu^+\mu^-$. The p_T^2 distribution of the muons produced by two photons at beam energies $\gtrsim 15$ GeV is plotted in Fig.4.7b. The distribution is well described by a power law $p_T^{-4.9\pm0.2}$, which is in good agreement with the QED results obtained with the Vermaseren program /38/. As we will discuss in Chap.8, one expects a similar behaviour for the p_T^2 distribution of jets produced in lowest order via the reaction

$$e^+e^- \rightarrow e^+e^-q\bar{q} \rightarrow e^+e^- + 2 \text{ jets.} \tag{4.08}$$

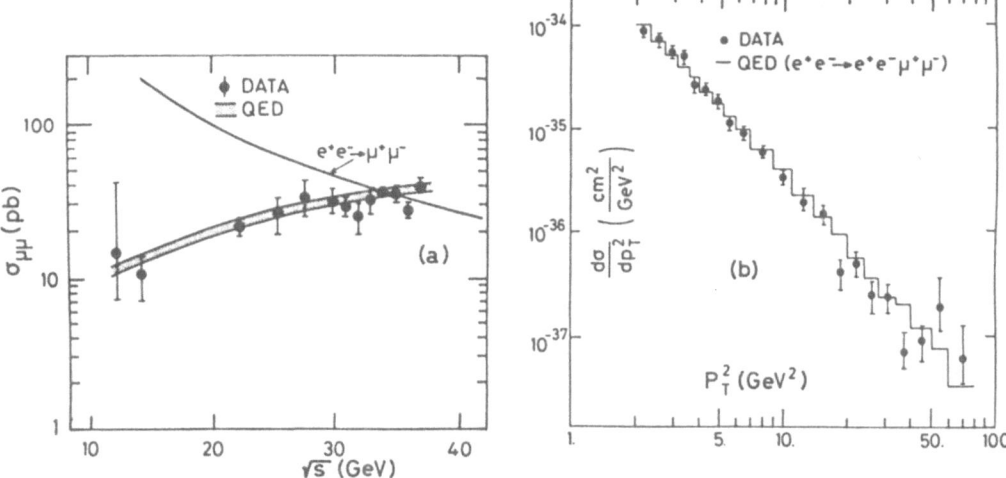

<u>Fig.4.7.</u>   Measurement of $e^+e^-\rightarrow e^+e^-\mu^+\mu^-$ (Mark J): a) Observed cross section for $e^+e^-\rightarrow e^+e^-\mu^+\mu^-$ at different e^+e^- energies compared to the QED prediction (shaded area) and to the process $e^+e^-\rightarrow\mu^+\mu^-$ (curve). b) Transverse momentum distribution of the muons.

The two-photon production of electron and muon pairs in the resonance region has been investigated by the PLUTO /43/, TASSO /44/ and Mark II /45/ groups in their analyses of two-photon resonances decaying into $\pi^+\pi^-$ (see Chap.6). Figure 4.8 shows the invariant mass spectrum for charged particle pairs (TASSO). Since the particles have not been identified, pion masses were assumed for all particles. Above the f region (M(f) = 1.27 GeV) the data agree at the 10% level with the QED prediction (using Vermaseren's program). The cross section for hadron pair production is expected to decrease much faster with $W_{\gamma\gamma}$ compared to QED reactions, so that above about 2 GeV the contribution from hadron pairs should be small. The PLUTO group analysed explicitly identified electron and muon pairs above 2 GeV for small Q^2 (no-tag) and large Q^2 ($0.1 < Q^2 <$ 10 GeV^2) and found agreement with the QED expectation within their 20 to 30% statistical error /46/.

The CELLO group studied lepton pair production in single-tag data at high Q^2 of the tagged photon ($<Q^2>$ = 9.5 GeV^2) /47/. The data have been compared to the Vermaseren program /38/ and to calculations based on the structure function formalism (see Chap.9). In the CELLO experiment the undetected scattered lepton is not restricted to small angles. It was found that in this case the variation of the four-momentum transfer of the untagged photon has to be properly taken into account, both in the target photon density (by differentiating (2.35) with respect to Θ_{2max}) and in the structure function. The results of such studies can be used for the measurement of the hadronic structure functions of the photon. The hadronic structure functions are predicted to be dominated by the Born graph for pointlike quark pair production and may thus behave similarly to the leptonic structure functions (see Chap.9).

The two-photon QED tests at PETRA have not yet reached the level where radiative corrections become important. Radiative corrections have been calculated in /48/ and found to be smaller than 10% in the single and double tag case and negligible in the no-tag case. In the CELLO analysis described above, the single-tag data have been corrected for initial state bremsstrahlung. The size of the correction is about 4%.

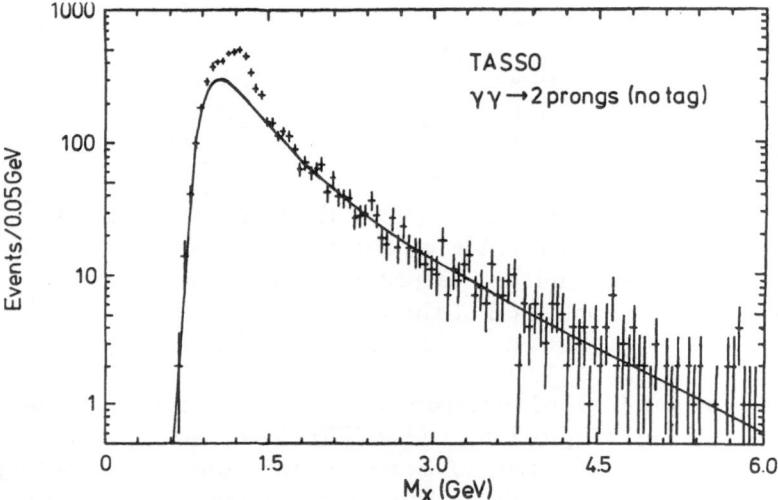

Fig.4.8.   Two-photon production of charged particle pairs (no-tag, TASSO): The invariant mass distribution of pairs of oppositely charged particles is compared to the QED prediction for lepton pair production.

5. Two-Photon Production of Meson and Baryon Pairs

In the following we discuss the non-resonant meson and baryon pair production by two photons. Results on resonances decaying into meson pairs, like $f \to \pi\pi$ and $f' \to K\bar{K}$, and on vector meson pair production near threshold will be presented in the next chapter (Chap.6).

At low energies one may compare the experimental results on charged hadron pair production with the expectation from the Born approximation, i.e. the QED coupling of the photons to pointlike hadrons. We will see that the continuum production of pairs of charged pions, even though pions are strongly interacting and not pointlike, can apparently be well described by the QED Born process up to the region of the f(1270) resonance. The analysis of resonances decaying into two charged mesons turns out to be very dependent on the proper understanding of the continuum (see Chap.6). In the presence of resonances the Born terms have to be modified by strong interactions. Therefore, in the resonance region no theoretical description of the continuum production of hadron pairs is possible with present knowledge.

Another theoretical approach has been employed at higher energies. Perturbative QCD calculations have been carried out which are expected to describe the cross section for hadron pairs produced with large momentum transfer, i.e. at high energies and large angles. These predictions have been tested for pion, kaon and proton pair production at $\gamma\gamma$ center of mass energies above 2 GeV.

5.1 Non-Resonant Meson Pair Production at Low Energies

At low energies the two-photon production of two charged mesons is expected to be enhanced compared to the production of two neutral mesons. Close to threshold charged meson pair production might be well described by the Born approximation given by the three graphs in Fig.5.1 /3/. These are the lowest order QED diagrams for the pointlike coupling of photons to charged (pseudo)scalar particles. They do not contribute to the production of two neutral scalar particles.

From these diagrams one obtains the differential cross section for scalar meson pair production by two real photons (ϑ^* = polar angle of the positive meson with respect to the incoming photons in the $\gamma\gamma$ center of mass system; β^* = velocity of the meson in the $\gamma\gamma$ center of mass system) /12, 36, 50/:

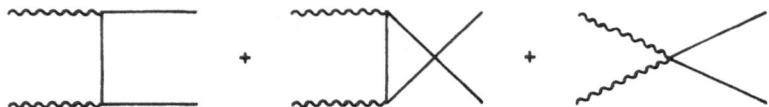

<u>Fig.5.1.</u>   Born diagrams for two-photon production of a pair of charged scalar particles.

$$d\sigma/d\cos\vartheta* = (\pi\cdot\alpha^2/W_{\gamma\gamma}^2)\cdot\beta*\cdot[(1-\beta^{*2})^2+\beta^{*4}\sin^4\vartheta*]/(1-\beta^{*2}\cos^2\vartheta*)^2 \qquad (5.01)$$

$$\lambda=0: \quad d\sigma/d\cos\vartheta* = (\pi\cdot\alpha^2/W_{\gamma\gamma}^2)\cdot\beta*\cdot(1-\beta^{*2})^2/(1-\beta^{*2}\cos^2\vartheta*)^2 \qquad (5.02)$$

$$\lambda=2: \quad d\sigma/d\cos\vartheta* = (\pi\cdot\alpha^2/W_{\gamma\gamma}^2)\cdot\beta*\cdot[\beta^{*2}\sin^2\vartheta*/(1-\beta^{*2}\cos^2\vartheta*)]^2. \qquad (5.03)$$

The cross section in (5.01) is averaged over the two possible $\gamma\gamma$ helicities $\lambda=0$ (both γ's with the same helicity) and $\lambda=\pm2$ (opposite γ helicities). The helicity contributions to (5.01) are given separately in (5.02) and (5.03). The differential cross section for pion pair production according to (5.01) is plotted in Fig.5.2 as a function of $\cos\vartheta*$ for different $W_{\gamma\gamma} = M(\pi\pi)$. The distribution is rather flat except for the forward direction, which is usually not within the experimental acceptance. The corresponding total cross sections, shown in Fig.5.3, are:

$$\sigma(\text{total}) = (\pi\cdot\alpha^2/W_{\gamma\gamma}^2)[2\beta*\cdot(2-\beta^{*2})-(1-\beta^{*4})\cdot\ln\{(1+\beta*)/(1-\beta*)\}] \qquad (5.04)$$

$$\sigma(\lambda=0) = (\pi\cdot\alpha^2/W_{\gamma\gamma}^2)[\beta*\cdot(1-\beta^{*2})+\tfrac{1}{2}\cdot(1-\beta^{*2})^2\cdot\ln\{(1+\beta*)/(1-\beta*)\}] \qquad (5.05)$$

$$\sigma(\lambda=2) = (\pi\cdot\alpha^2/W_{\gamma\gamma}^2)[\beta*\cdot(3-\beta^{*2})+\tfrac{1}{2}\cdot(\beta^{*4}+2\beta^{*2}-3)\cdot\ln\{(1+\beta*)/(1-\beta*)\}]. \qquad (5.06)$$

The total cross section for $e^+e^-\rightarrow e^+e^-\pi^+\pi^-$ using the Born cross section (5.04) and the equivalent photon approximation is shown as a function of the beam energy in Fig.4.3.

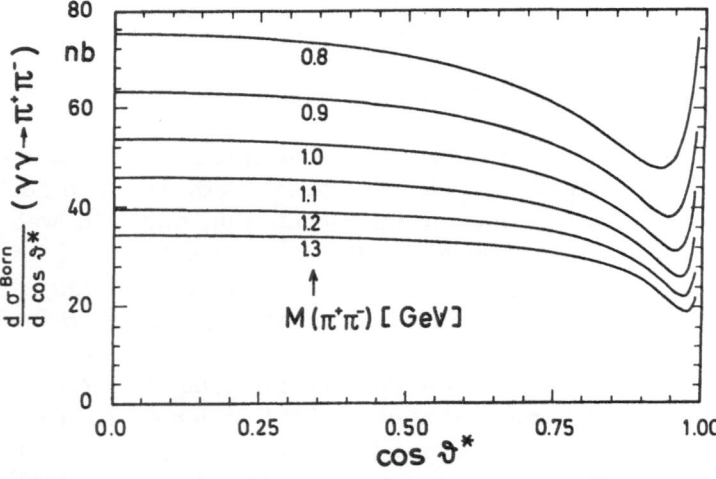

<u>Fig.5.2.</u>   Angular distribution for $\gamma\gamma\rightarrow\pi^+\pi^-$ (Born approximation)

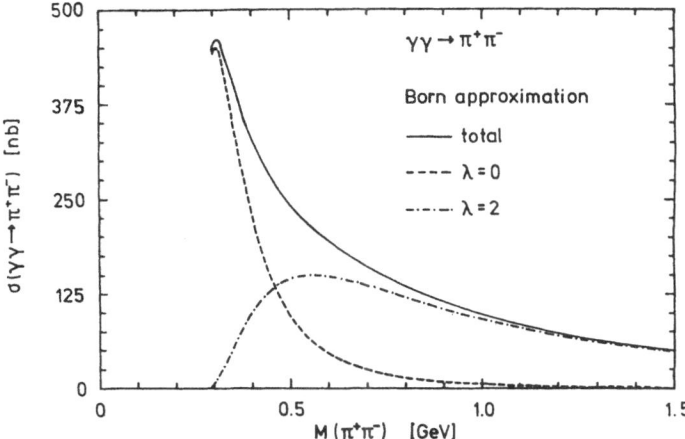

<u>Fig.5.3.</u>   Cross section for $\gamma\gamma\to\pi^+\pi^-$ (Born approximation). The curves show the cross sections for $\gamma\gamma$ helicities 0, 2 and the sum of both.

For a comparison with experiments, the curves in Fig.5.2 and Fig.5.3 can only serve as a guide line, since one does not know a priori to what extent the Born approximation is applicable. Since the hadrons are not pointlike the approximation has to break down with increasing energy; strong inter-action effects like final state interactions, the opening of other channels and additional exchange contributions in the t channel have to be taken into account. However, it has been argued /12/ that for the $\pi\pi$ partial waves with $J\geqq2$ (which includes the whole $\lambda=2$ contribution) the Born ap-proximation should be good up to ~1 GeV, because the corresponding $\pi\pi$ phase shifts are small. The connection between the elastic $\pi\pi$ phase shifts and the phase of $\gamma\gamma\to\pi\pi$ is given by the unitarity condition which requires that both phases (for the same isospin) are the same if the contributions from other inelastic channels are small (Watson theorem /51/). Note how-ever, that the modulus of the amplitude for $\gamma\gamma\to\pi\pi$ is independent of the phase shift, in contrast to the elastic channel. That allows having a large cross section even if the phase shift is small, as for the Born approxi-mation near threshold.

In the case of kaon pair production, the Born approximation may not be a good description of the cross section even very close to threshold, be-cause the KK phase shift is large near threshold.

An accurate description of the $\pi\pi$ continuum turns out to be crucial for the determination of the $\gamma\gamma$ width of the f, which decays mainly into $\pi\pi$. Attempts have been made to improve the theoretical description of the $\pi\pi$ and $K\bar{K}$ pair production beyond the Born approximation, especially to include resonances in a reasonable way /49, 50/. In the model of Mennessier /50/ a coupled channel treatment of $\pi\pi$ and $K\bar{K}$ is pursued. The pion exchange (Fig.5.1) as well as the ρ and ω exchanges are unitarized by including $\pi\pi$ and $K\bar{K}$ final state interactions as given by the measured phase shifts (Fig.5.4). In addition, a direct coupling of the photons to res-

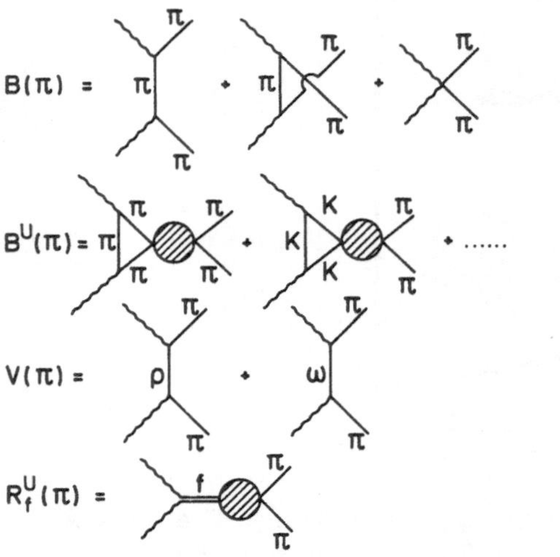

Fig.5.4. The diagrams contributing to the reaction $\gamma\gamma \to \pi^+\pi^-$ according to the Mennessier model /50/. B, V and R are the contributions from the Born approximation, vector meson exchange and direct resonance production, respectively. The superscript u denotes unitarized amplitudes.

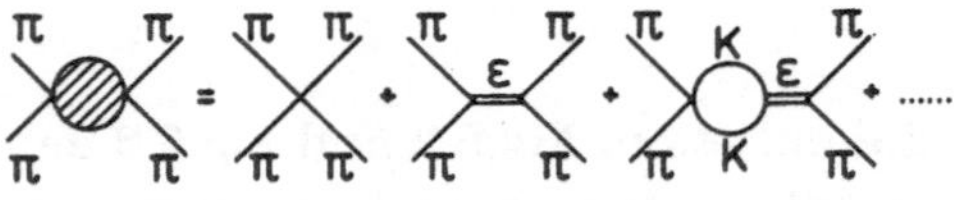

onances is introduced. With such an approach, one can hope to get from $\gamma\gamma$ scattering information about the $\pi\pi$ and $K\bar{K}$ phase shifts. Especially interesting is the low $\pi\pi$ mass region, where the possible existence of a scalar resonance is still subject to experimental investigations.

The process $\gamma\gamma \to \pi^+\pi^-$ has been measured at the DCI storage ring in the $\pi\pi$ mass range from threshold to about 0.7 GeV /41/. The pion pairs have been separated from the electron and muon pair background by determining the particle mass from a kinematical reconstruction of the event (see Fig.4.6 in Chap.4). With such a method for $\pi\pi$ detection near threshold, this experiment is superior to experiments at PETRA or PEP. Unfortunately the experiment suffers from lack of statistics. The measured cross section (within the acceptance) of $\sigma(\gamma\gamma \to \pi^+\pi^-) = 69\pm15$ pb has to be compared to 34 pb for the Born approximation. The model of /50/ is able to describe the data by including a scalar resonance $\varepsilon(800)$ (see also Sect.6.5). However, the effect has a significance of only 2.3 standard deviations, not enough to establish a resonance in this region. A measurement of the MARK II group, covering $\pi^+\pi^-$ masses down to 0.5 GeV, is consistent with the assumption that only the Born term contributes below about 1 GeV (see Fig.6.15).

5.2 Hadron Pair Production above 2 GeV

For high energies, hadron pair production has been calculated using the techniques of perturbative QCD. Brodsky and Lepage /52/ have developed a model for calculating two-photon pair production cross sections at large transverse momenta. In this model it is assumed that for scattering processes involving high momentum transfers the scattering amplitude can be factorized into two parts: a hard scattering amplitude T_H and a distribution function Φ for the valence quarks in the hadron (see Fig.5.5):

$$M \sim \Phi^* \cdot T_H \cdot \Phi. \tag{5.07}$$

The soft part, the parton distribution function Φ, cannot be calculated. One has to make reasonable assumptions about the form of the parton distribution. The absolute normalization of Φ is then obtained by comparing with other processes containing the same distribution function. The hard scattering amplitude T_H is calculated perturbatively by summing up the lowest order QCD diagrams (Fig.5.5).

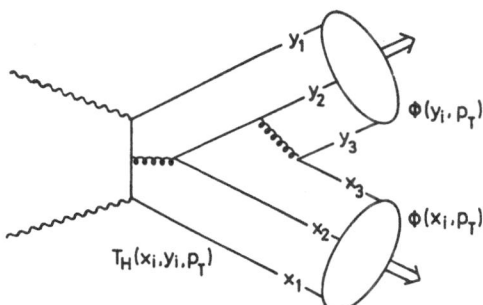

Fig.5.5.   Illustration of the QCD evaluation of two-photon production of hadron pairs taking as example $\gamma\gamma \to p\bar{p}$.

From the dimensional counting rule (i.e. counting the elementary fields involved in a subprocess) the following energy dependences of the differential cross section at fixed angle are expected for meson and baryon pair production:

$$d\sigma(\gamma\gamma \to M\bar{M})/dt \sim s^{-4} \qquad d\sigma(\gamma\gamma \to B\bar{B})/dt \sim s^{-6} \tag{5.08}$$

(s, t are the usual Mandelstam variables for the process $\gamma\gamma \to M\bar{M}$, $B\bar{B}$). The pointlike coupling, as assumed in the Born approximation, yields an s^{-2} dependence. Another feature predicted from the QCD calculations is the peaking of the angular distribution of the produced pairs in the forward direction. However, the calculations are most reliable at large scattering angles, i.e. where the momentum transfer is high.

5.2.1   A Measurement of Hadron Pair Production above 2 GeV

The PLUTO group has looked for charged hadron pairs with invariant masses above 2 GeV, without separating pions, kaons and protons /53/. An earlier analysis yielded upper limits /46/. The new analysis uses data from an integrated luminosity of $\sim$40 pb^{-1}. The charged hadron pairs have been selected as those pairs which have not been positively identified as electrons or muons. Since the QED processes are by far dominant, this procedure requires an excellent electron and muon rejection capability. That has been achieved using the barrel shower counters and the muon detection system of the PLUTO detector. Events with two oppositely charged tracks with |cosϑ|<0.6 (ϑ is the laboratory polar angle with respect to the beam) and momenta larger than 0.9 GeV were selected. The photons were restricted to have low Q^2 values by the requirement that there was no tag in the small and large angle taggers. Exclusively produced pairs were selected by demanding that $|\sum\vec{p}_T|$<0.5 GeV and that the tracks were coplanar with the beam within 6°. The process $\gamma\gamma\rightarrow e^+e^-$ was rejected by a cut in the shower energies, leading to about 20% loss of hadron pairs but virtually no remaining background from electrons. The remaining 987 events come mainly from the reaction $\gamma\gamma\rightarrow\mu^+\mu^-$ with a small contribution from hadron pair production.

For the separation of the hadron pairs from the muons the momenta and angles of the particles were restricted to a range in which muons are identified with a probability greater than 98%. 651 events fulfil this requirement. Only 16 of them were not identified as $\mu^+\mu^-$. After subtraction of the estimated $\mu^+\mu^-$ background, 15.1 hadron pair events are left yielding the ratio of hadron pair production to muon pair production:

$$\frac{\sigma(e^+e^-\rightarrow e^+e^-h^+h^-)}{\sigma(e^+e^-\rightarrow e^+e^-\mu^+\mu^-)}\ (W_{\pi\pi}>2.0\ \text{GeV})\ =\ 0.042\ \pm\ 0.013(\text{stat.})\ \pm\ 0.008(\text{syst.}).$$

This ratio is more than an order of magnitude smaller than expected for a production of pointlike hadrons, but it is in reasonable agreement with the QCD calculations of /52/, which predict for this ratio 0.020 ± 0.001. In Fig.5.6 the ratio of the two-photon produced hadron pairs to muon pairs is plotted versus the mass of the hadron pair, assuming pion masses for all hadrons. The lowest $\pi\pi$ mass of 2 GeV corresponds to a KK mass of 2.21 GeV and to a p$\bar{\text{p}}$ mass of 2.73 GeV. The QCD calculation suggests that the hadron

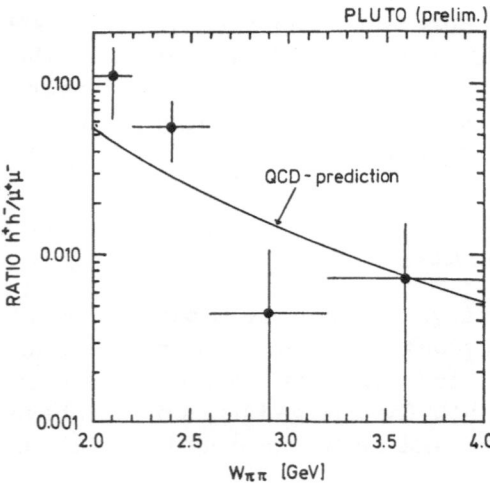

Fig.5.6. Measurement of two-photon production of charged hadron pairs: Ratio of charged hadron pair production to muon pair production as a function of the invariant mass of the pair assuming all hadrons to be pions (no-tag, PLUTO).

composition within the acceptance is 48% $\pi^+\pi^-$, 48% K^+K^- and 4% $p\bar{p}$. The QCD curve in Fig.5.6 was obtained with this composition assuming, as in the data, pion masses for all particles.

5.2.2 Proton-Antiproton Pair Production

Proton-antiproton production by two photons

$$\gamma\gamma \rightarrow p\bar{p} \tag{5.09}$$

was first seen by the TASSO collaboration /54/. Eight events were observed in the $p\bar{p}$ mass range from 2.0 to 2.6 GeV. This analysis has been repeated with much higher integrated luminosity ($\sim$74 pb^{-1}) increasing the number of events by an order of magnitude /55/.

Candidate events for the reaction $\gamma\gamma\rightarrow p\bar{p}$ have been selected from events with two oppositely charged tracks in a polar angular range $|\cos\vartheta|<0.8$ and in a momentum range $0.35\leq p\leq1.6$ GeV. The $p\bar{p}$ pairs have been identified by means of the time-of-flight (TOF) information of the counters surrounding the TASSO central drift chamber. In Fig.5.7 the square of the mass calculated from TOF for the positive track is plotted versus the same quantity for the negative track. The cluster of $p\bar{p}$ events is clearly separated (see also Fig.3.7 in Chap.3). The final sample contains 72 $p\bar{p}$ events with invariant masses between 2.0 and 3.1 GeV and with $|\sum\vec{p}_T|$ < 0.1 GeV. The $|\sum\vec{p}_T|^2$ distribution is consistent with the assumption that the observed $p\bar{p}$ events are exclusively produced in the reaction (5.09). The background contributes less than 5 events.

In Fig.5.8 the differential cross section for $\gamma\gamma\rightarrow p\bar{p}$ is given for two $p\bar{p}$ mass intervals as a function of $|\cos\vartheta^*|$ (ϑ^* = polar angle of the p in the $\gamma\gamma$ rest system). In both intervals the differential cross section is consistent with a flat $\cos\vartheta^*$ dependence. The covered angular range $|\cos\vartheta^*|<0.6$ corresponds to momentum transfers between the protons from 0.7 to 2.5 GeV^2. The measured differential cross sections are compared to the Born

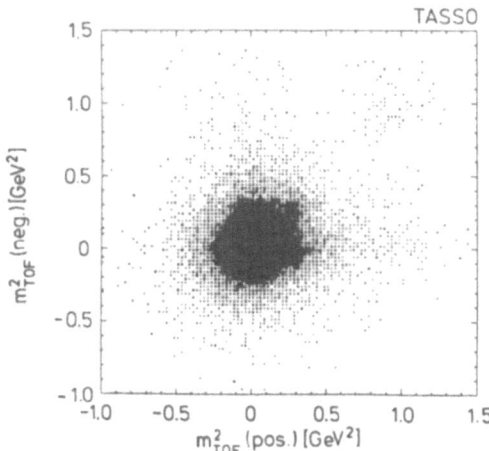

Fig.5.7.   Identification of $p\bar{p}$ production by time-of-flight (TOF): Square of the mass calculated from TOF for the negative track versus the same quantity for the positive track in events with 2 charged tracks detected (no-tag, TASSO).

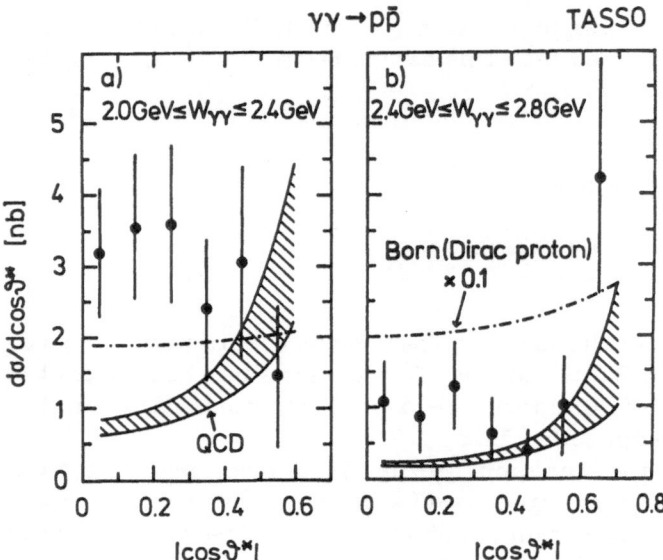

<u>Fig.5.8.</u>   The differential cross section $d\sigma(\gamma\gamma\to p\bar{p})/d\cos\vartheta^*$ (TASSO): a) $2.0<W_{\gamma\gamma}<2.4$ GeV and b) $2.4<W_{\gamma\gamma}<3.0$ GeV. Data are compared to the Born approximation for a Dirac proton (dashed-dotted curve, scaled by 1/10) and to QCD calculations /58/ (hatched area). The band for the QCD calculations corresponds to different choices for the parton distribution in the protons.

approximation for the process $\gamma\gamma\to p\bar{p}$, treating the protons and antiprotons as pointlike fermions with a QED coupling to the photons (dashed-dotted curve, scaled by 1/10). This approximation gives cross sections which are roughly an order of magnitude too large. Including the anomalous magnetic moment of the proton yields even higher cross sections /56/. The cross section for $\gamma\gamma\to p\bar{p}$, integrated over the angular range $|\cos\vartheta^*|<0.6$, is plotted in Fig.5.9 as a function of the $p\bar{p}$ mass. Using the vector meson dominance model, the expectation for $\sigma(\gamma\gamma\to p\bar{p})$ was calculated from measured data of the inverse reactions $p\bar{p}\to\rho^\circ\rho^\circ$ and $p\bar{p}\to\rho^\circ\omega$ /57/. For $|\cos\vartheta^*|<0.6$ a cross section of 0.1 to 0.6 nb was obtained.

The results are also compared to a calculation using perturbative QCD methods as described above /58/. In this case the 'soft' parton distribution function Φ in (5.07) has been normalized to the process $J/\psi\to p\bar{p}$. Around $\cos\vartheta^*=0$, where the four-momentum transfer is largest, the rate predicted by QCD is about a factor of 3 to 5 below the measured data (Fig.5.8). QCD predicts a strong rise of the cross section in the forward direction. Given the limited acceptance ($|\cos\vartheta^*| \leq 0.6$) and the low statistics, the data are not yet sensitive enough to see such a behaviour.

In conclusion, the production of pion, kaon and proton pairs by two photons above 2 GeV is clearly at least an order of magnitude smaller than expected from a pointlike coupling. QCD calculations give the correct or-

der of magnitude for the sum of hadron pair production. The data show a
tendency to approach the prediction from above as $W_{\gamma\gamma}$ increases. Unfortunately, in the region where the QCD calculations become safer few events
are found.

Recently, the absolute normalization of the QCD calculation for $\gamma\gamma \rightarrow p\bar{p}$
/58/ has been questioned. In another calculation of this process /59/ a
cross section about a factor of 100 lower was obtained. The problem is still
under investigation at the time this review is written /60/.

If the calculations of /59/ are correct, one has to explain the abundant
$p\bar{p}$ production, e.g. one may wonder if resonances contribute. Independent
of the normalization, the lowest order QCD diagrams lead to a $\Delta^{++}\overline{\Delta^{++}}$ production which is more than an order of magnitude larger than $p\bar{p}$ production /59/. Preliminary data from TASSO show that this is not the case
for baryon pair masses around 3 GeV /61/.

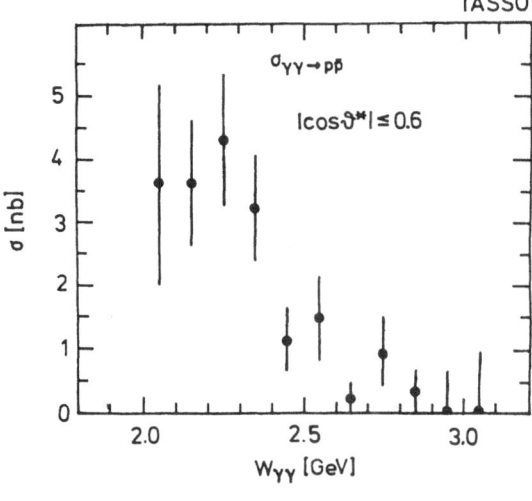

Fig.5.9.   The   cross   section
$\sigma(\gamma\gamma \rightarrow p\bar{p})$ for $|\cos\vartheta^*| < 0.6$ as   a
function of the $p\bar{p}$ mass (TASSO).

6. The Coupling of Two Photons to Meson Resonances

6.1 Introduction

In the last few years the study of two photon resonances has been particularly prolific and many more interesting experimental results are expected to become available in the near future. In 1960, F.Low suggested measuring the $\gamma\gamma$ width of the π° at e^+e^- storage rings /2/, but it took nearly two decades before the first resonance produced by two photons was observed at a storage ring. In 1979, the Mark II collaboration reported the measurement of the $\gamma\gamma$ width of the η' at SPEAR /62/. Then, within only a few years, the $\gamma\gamma$ couplings of the tensor mesons f, A_2 and f' had been determined at PETRA and SPEAR. Thus the $\gamma\gamma$ widths of all flavour-neutral members of the pseudoscalar and tensor meson multiplets are known by now (Tables 6.1,2), allowing interesting theoretical tests, in particular with respect to SU(3) symmetry.

The two-photon production of mesons has some advantages compared to meson production in hadronic processes. First, the production rate can be absolutely calculated from QED, with the $\gamma\gamma$ width as the only free parameter. Secondly, in two-photon reactions the meson can be produced alone, i.e. without additional hadronic debris. Thirdly, the study of resonances is in some cases easier, because the quantum numbers of the final state are restricted in two-photon reactions. An obvious restriction is that meson resonances which couple to two photons must have charge conjugation C=+1. Furthermore, systems with spin J=1 do not couple to two real photons according to Yang's theorem /63/. In principle, there is the possibility of measuring the coupling of off-shell photons to axial-vector mesons, but in most of the experiments discussed here the kinematical situation is such that spin J=1 contributions can be neglected. For the resonances with even C parity, the G parity depends only on the isospin I of the resonance: $G=C\cdot(-1)^I$. Thus isoscalars will decay into an even number of pions and isovectors into an odd number of pions.

The cross section for the production of a resonance R with spin J by two real photons can be parametrized by a Breit-Wigner function /12/:

$$\sigma(\gamma\gamma \to R) = 8\pi(2J+1) \frac{\Gamma \cdot \Gamma_{\gamma\gamma}}{(W_{\gamma\gamma}^2 - M_R^2)^2 + \Gamma^2 M_R^2} \qquad (6.01)$$

Table 6.1. Measured two-photon widths of pseudoscalar and tensor meson resonances: The results for the tensor mesons were obtained assuming pure $\gamma\gamma$ helicity 2. If two errors are given, the first is statistical and the second systematic. For the f' the 4th column contains the product of the $\gamma\gamma$ width and the branching ratio into $K\bar{K}$.

R	Reference		$\Gamma_{\gamma\gamma}$ [keV]	Remarks
π^0	PDG	/72/	$(7.85 \pm 0.54)\cdot 10^{-3}$	
η	PDG	/72/	0.324 ± 0.046	Primakoff
	C.Ball	/35/	$0.56 \pm 0.12 \pm 0.10$	
η'		/76/	5.4 ± 2.1	πp scatt.
	Mark II	/62/	$5.8 \pm 1.1 \pm 1.2$	
	CELLO	/78/	$6.2 \pm 1.1 \pm 0.8$	
	JADE	/77/	$5.0 \pm 0.5 \pm 0.9$	
	TASSO	/80/	$4.1 \pm 0.4 \pm 1.5$	prelim.
	PLUTO	/81/	$3.7 \pm 0.2 \pm 0.6$	prelim.
f	Pluto	/43/	$2.3 \pm 0.5 \pm 0.3$	
	TASSO	/44/	$3.2 \pm 0.2 \pm 0.6$	
	Mark II	/45/	$3.6 \pm 0.3 \pm 0.5$	
	C.Ball	/108/	$2.7 \pm 0.2 \pm 0.6$	
	CELLO	/81/	$2.5 \pm 0.1 \pm 0.5$	prelim.
	JADE	/80/	$2.3 \pm 0.2 \pm 0.5$	prelim.
	Mark II	/81/	$2.5 \pm 0.1 \pm 0.4$	prelim.
A_2	C.Ball	/108/	$0.77 \pm 0.18 \pm 0.27$	
	CELLO	/78/	$0.81 \pm 0.19 \pm 0.27$	
	JADE	/80/	$0.84 \pm 0.07 \pm 0.15$	prelim.
f'	TASSO	/110/	$0.11 \pm 0.02 \pm 0.04$	$f'\to K\bar{K}$

where Γ is the total width, M_R the mass and $\Gamma_{\gamma\gamma}$ the $\gamma\gamma$ width of the resonance. For a narrow resonance this formula becomes:

$$\sigma(\gamma\gamma\to R) = 8\pi^2(2J+1)\, \frac{\Gamma_{\gamma\gamma}}{M_R}\, \delta(W_{\gamma\gamma}^2 - M_R^2). \qquad (6.02)$$

The photon flux is usually calculated as given in /15/ including only transverse photons. The equivalent photon approximation is conveniently used to get an estimate for the production rate of a resonance R in the process $e^+e^-\to e^+e^-R$ /2/:

$$\sigma(e^+e^-\to e^+e^-R) \approx 16\alpha^2(\log E/m_e)^2\, f(M_R/2E)\, \frac{(2J+1)\Gamma_{\gamma\gamma}}{M_R^3} \qquad (6.03)$$

with $f(x) = (2+x^2)^2 \log(1/x) - (1-x^2)(3+x^2)$.

Table 6.2.   Compilation of $\gamma\gamma$ widths of resonances: The table contains average values and upper limits (c.l. = confidence level) for the product of the $\gamma\gamma$ width and the decay branching ratio into the final state X (2nd col.). Note that X is not necessarily the observed decay mode. The average values are obtained from the published results only (see Table 6.1).

R	X	$\Gamma(R{\to}\gamma\gamma)\cdot Br(R{\to}X)$ [keV]		Remarks	Ref.	Exp.
π°	all	$(7.85 \pm 0.54)\cdot 10^{-3}$			/72/	
η	all	0.343 ± 0.044				
η'	all	5.5 ± 0.7				
f	all	2.95 ± 0.30				
A_2	all	0.79 ± 0.23				
f'	$K\bar{K}$	0.11 ± 0.05			/110/	TASSO
S*	$\pi\pi$	< 0.8	95% c.l.		/108/	C.Ball
	all	< 0.8	95% c.l.	prelim.	/80/	JADE
ε	$\pi^+\pi^-$	< 1.5	95% c.l.	1.3<M<1.5 GeV	/44/	TASSO
ι	$\rho^\circ\rho^\circ$	< 1.0	95% c.l.		/114/	TASSO
	$K\bar{K}\pi$	< 8.0	95% c.l.		/24/	Mk II
	$K\bar{K}\pi$	< 7.0	95% c.l.	prelim.	/80/	TASSO
Θ	$\rho^\circ\rho^\circ$	< 1.2	95% c.l.		/114/	TASSO
	$K\bar{K}$	< 0.3	95% c.l.		/110/	TASSO
	$K\bar{K}$	< 0.4	95% c.l.		/116/	Mk II
	$\eta\eta$	< 0.3	95% c.l.		/115/	C.Ball
η_c	all	< 20	90% c.l.	see Table 6.4	/129/	C.Ball
χ_0	all	< 10	90% c.l.	prelim.	/134/	C.Ball
χ_2	all	1.8 ± 1.3		prelim.	/134/	C.Ball
?(2100)	$2\pi^+2\pi^-$	$1.25 \pm 0.5 \pm 0.5$		prelim.	/152/	TASSO

Figure 6.1 shows the expected cross section for 17 GeV and 50 GeV beam energy. Higher energies do not dramatically increase the cross sections. On the other hand the detection efficiencies decrease at higher beam energies because the boosts of the $\gamma\gamma$ system are on average larger.

Since photons are not directly involved in strong interactions, they are well suited as probes to determine the electric charge content of a hadron and thus its quark composition. Therefore, measurements of the two photon couplings of C-even mesons, as well as radiative transitions from C-odd mesons, are especially useful in studying the nature of the hadrons.

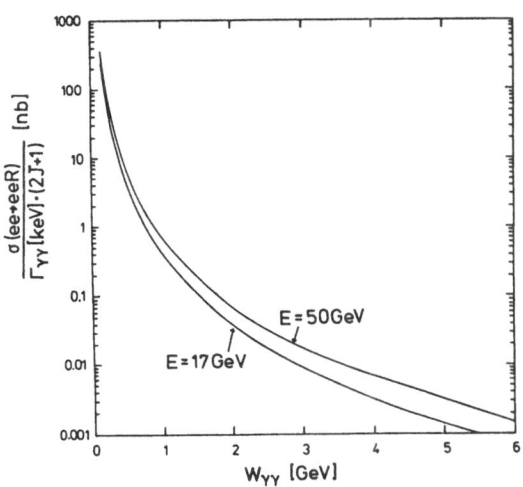

Fig.6.1. The e^+e^- cross section for the production of a resonance via two photons.

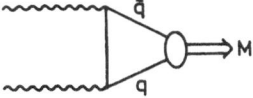

Fig.6.2. Coupling of two photons to a meson according to the quark model.

In the quark model two photons couple to a meson via the charges of the quarks in the meson according to Fig.6.2. A meson is in general a coherent mixture of $q\bar{q}$ pairs with different flavours. The flavour dependence of the quark charges makes the $\gamma\gamma$ width particularly sensitive to the size and the relative phases of the flavour components in this mixture.

Since they depend on the fourth power of the quark charges, the $\gamma\gamma$ widths of resonances should be well suited to determine the charge of the quarks and in particular to distinguish between models with integral and fractional quark charges. In models which do not have the conventional fractional charge assignment, the charge of a quark with a given flavour depends on its colour. Thus in these models the electromagnetic current is not necessarily a colour singlet but may have an octet component. The colour singlet photon always sees the charges averaged over the colours, which are, independent of the model, the same as the fractional Gell-Mann/Zweig charges. Only the colour octet photon measures the actual charges. Below the threshold for open colour the colour octet cannot contribute in processes involving only one photon, whereas in two-photon reactions two colour octets can couple to a colour singlet final state. The implications of the integrally charged quark model will be discussed in more detail in connection with the $\gamma\gamma$ width of the pseudoscalars (Sect.6.3.5).

Mesons may not only be made of quarks and antiquarks. If QCD is the correct theory to describe strong interactions, its properties as a nonabelian gauge theory suggest that there exist bound states of gluons, the field quanta of QCD /64/. These states, called gluonia or glueballs, are composed of two or more gluons, bound in specific configurations with well defined spin and parity assignments. With respect to our subject of two-photon physics, gluonia are characterized by the fact that they do not contain charged valence constituents. Thus the two-photon widths are expected to be suppressed. Mixing with flavour neutral quark states of the same spin and parity can complicate the situation. In principle, the mixing matrix can be determined from the measured $\gamma\gamma$ widths, if the mixed states have been clearly identified. But as yet, the existence of gluonium states has not been proven and one needs much more information for the construction of a consistent model for the mesons.

The comparison of the $\gamma\gamma$ width of a meson to its production rate in radiative J/ψ decays provides a possible handle in deciding about its quark or gluon nature. In radiative J/ψ decays, i.e. $J/\psi \rightarrow \gamma X$, where X is a hadronic system, the transition of J/ψ to hadrons proceeds in lowest order perturbative QCD via two gluons (Fig.6.3a). Therefore this channel seems to be favourable for the observation of bound states of two gluons. Since photons as well as gluons are massless spin-1 particles, the structure of the two-photon and the two-gluon coupling is similar, with the difference that the photons couple to electric charges and the gluons to colour charges. Thus, comparing these two processes provides the information on the charge and colour or, equivalently, on the quark and gluon composition of the meson.

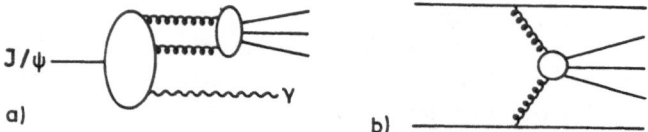

Fig.6.3.   Two-gluon coupling to hadrons: a) Radiative decay of the J/ψ, b) doubly peripheral hadron-hadron scattering via two-gluon interaction.

Another "glueball favoured" channel, which can be related to $\gamma\gamma$ scattering, is the so-called pomeron-pomeron scattering observed in hadronic interactions /see e.g. 65/. This process may be dominated by gluon-gluon scattering according to the diagram in Fig.6.3b.

In addition to the conventional quark resonances and possible gluonic bound states, also multiquark states such as $qq\bar{q}\bar{q}$ may exist /66/. These states are expected to be very broad, since they can fall apart in normal $q\bar{q}$ states, and therefore they should be hard to detect. In cases where the four-quark state has a mass smaller than the threshold for the decay into its 'fall apart' components, it could have a width comparable to normal hadrons. Four-quark configurations with 'fall apart' pieces having the quantum numbers of the photon could have relatively large $\gamma\gamma$ widths. As yet, no four-quark states have been firmly identified, but it has been suggested that they may explain some features of the observed low-lying 0^+ mesons /67/ and that the large cross section observed in the two-photon production of $\rho°\rho°$ could be due to a combination of four-quark states (see Sect.6.7). Again, we have to consider mixing with other states of the same quantum numbers. Thus in the most general case a rather complicated mixing scheme between quark states, gluonia and four-quark states could develop.

The typical problems of light quark spectroscopy, which may be caused by yet-to-be understood non-perturbative QCD effects, seem to be absent in heavy quark systems, for which perturbative QCD has been successfully applied. According to perturbative QCD the hadronic width of flavour neutral states is dominated by the decay via the minimum number of gluons. That means that the C-even states which couple to two photons will decay into hadrons via two gluons. Thus the ratio of the hadronic width to the two photon width is proportional to the square of the strong coupling constant α_s /68/.

This chapter on $\gamma\gamma$ resonances is organized as follows: We start with a brief introduction to the quark model description of $\gamma\gamma$ couplings of meson resonances. Next the experimental results on the $\gamma\gamma$ couplings of the pseudoscalar and tensor mesons are reviewed, followed by a discussion of the theoretical implications of these results. Then we summarize what is known and what is not known about $\gamma\gamma$ couplings of scalar mesons and charmonium states. Finally we report on the search for new resonances in the four pion final state and on the production of vector meson pairs.

6.2 Meson Resonances in the Quark Model

In the following we want to summarize SU(3) relations for the $\gamma\gamma$ coupling of mesons. The $\gamma\gamma$ width can be expressed in terms of coupling constants $g_{M\gamma\gamma}$ /see e.g. 69/. For pseudoscalars and scalar resonances one defines:

$$\Gamma_{P\gamma\gamma} = \frac{m_P{}^3}{64\pi}\, g_{P\gamma\gamma}{}^2 \tag{6.04}$$

$$\Gamma_{S\gamma\gamma} = \frac{m_S{}^3}{16\pi}\, g_{S\gamma\gamma}{}^2. \tag{6.05}$$

The tensor mesons have in general two couplings corresponding to the two possible $\gamma\gamma$ helicities (the two photons with equal and opposite helicity).

In the quark model the mesons are represented by coherent mixtures of $q\bar{q}$ pairs with different flavours:

$$|M> = \sum_q c_q |q\bar{q}>. \tag{6.06}$$

The coupling of two photons (with a given $\gamma\gamma$ helicity) to a quark pair is proportional to the square of the quark charges:

$$<q\bar{q}|\gamma\gamma> \sim e_q{}^2 \cdot \Psi_q(0) \qquad \text{(S wave)}$$
$$<q\bar{q}|\gamma\gamma> \sim e_q{}^2 \cdot \Psi_q{}'(0) \qquad \text{(P wave)} \tag{6.07}$$

where $\Psi_q(0)$ is the radial quark wave function at the origin. For quark pairs in a P wave (e.g. scalar and tensor resonances) the wave function vanishes at the origin and Ψ_q has to be replaced by the first derivative of Ψ_q at zero. If $\Psi_q(0)$ is independent of the quark flavour the $\gamma\gamma$ coupling constant of a meson M can be related to the quark charges using (6.06) and (6.07):

$$g_{M\gamma\gamma} \sim <M|\gamma\gamma> \sim \sum_q c_q \cdot e_q^2 = <e_q^2>. \tag{6.08}$$

If SU(3) symmetry holds the coefficients c_q are given by the SU(3) representations, which are for the flavour neutral members of the pseudoscalar (tensor) meson nonet:

$$\begin{aligned}
\pi^0 \ (A_2) &= (d\bar{d}-u\bar{u})/\sqrt{2} \\
\eta_8 \ (f_8) &= (u\bar{u}+d\bar{d}-2s\bar{s})/\sqrt{6} \\
\eta_1 \ (f_1) &= (u\bar{u}+d\bar{d}+s\bar{s})/\sqrt{3}.
\end{aligned} \tag{6.09}$$

The indices 8 and 1 denote the flavour octet and flavour singlet isoscalars. For these states, the effective squared charge defined in (6.08) is (v denotes the isovector, i.e. π^0 or A_2):

$$\begin{aligned}
<e_q^2>_v &= (e_d^2-e_u^2)/\sqrt{2} = -1/(3\sqrt{2}) \\
<e_q^2>_8 &= (e_u^2+e_d^2-2e_s^2)/\sqrt{6} = 1/(3\sqrt{6}) \\
<e_q^2>_1 &= (e_u^2+e_d^2+e_s^2)/\sqrt{3} = 2/(3\sqrt{3}).
\end{aligned} \tag{6.10}$$

Since SU(3) symmetry is broken by the large mass of the strange quark, the observable isoscalars η and η' (f' and f) are mixtures of the SU(3) singlet and octet states (we neglect in the following discussion admixtures of heavy quarks like charm quarks):

$$\begin{aligned}
\eta &= \cos\Theta \ \eta_8 - \sin\Theta \ \eta_1 \\
\eta' &= \sin\Theta \ \eta_8 + \cos\Theta \ \eta_1 \\[4pt]
f' &= \cos\Theta \ f_8 - \sin\Theta \ f_1 \\
f &= \sin\Theta \ f_8 + \cos\Theta \ f_1.
\end{aligned} \tag{6.11}$$

The case that the mesons are either made of the light u, d quarks or of s quarks only, is called ideal mixing, corresponding to $\Theta = 35.3°$. Neglecting any possible mass dependence the ratios of the coupling constants within a SU(3) nonet depend only on the quark charges and on the mixing angle:

$$\begin{aligned}
g_{\pi\gamma\gamma} &: g_{\eta\gamma\gamma} : g_{\eta'\gamma\gamma} = g_{A2\gamma\gamma} : g_{f'\gamma\gamma} : g_{f\gamma\gamma} = \\
&<e_q^2>_v : \cos\Theta<e_q^2>_8-\sin\Theta<e_q^2>_1 : \sin\Theta<e_q^2>_8+\cos\Theta<e_q^2>_1 = \\
&-\sqrt{3} : \cos\Theta-2\sqrt{2}\sin\Theta : \sin\Theta+2\sqrt{2}\cos\Theta.
\end{aligned} \tag{6.12}$$

A possible mass dependence of the $g_{M\gamma\gamma}$ is strongly model dependent. For example, in a nonrelativistic quark model, it depends on the radial shape of the potential for the $q\bar{q}$ bound state (see e.g. /70/).

6.3 The Pseudoscalar Mesons

6.3.1 The Two-Photon Coupling of the π^0

In the first theoretical paper on two-photon physics at e^+e^- storage rings the measurement of the $\gamma\gamma$ width of the π^0 was suggested /2/. The $\gamma\gamma$ decay of the π^0 played a central role in establishing the colour degree of freedom of quarks and thus became a cornerstone for the construction of the colour gauge theory, quantum chromodynamics. The $\gamma\gamma$ width of the π^0 can be calculated rigorously /71/ using the concept of partially conserved axial vector currents together with the axial anomaly which is connected to the triangle diagram in Fig.6.4.

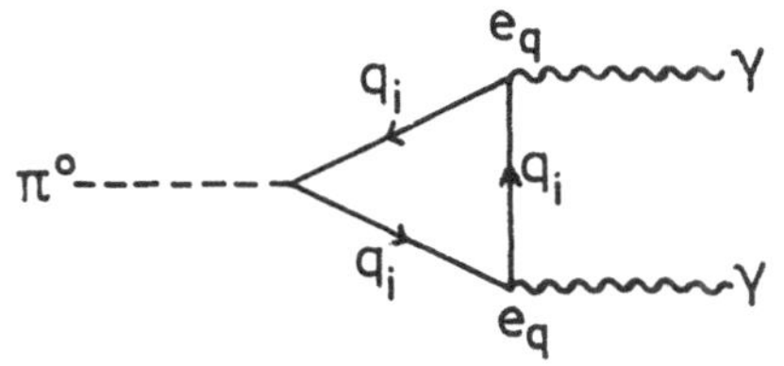

Fig.6.4.   Triangle diagram of the π^0 decay.

In the soft pion limit, $m_{\pi^0}=0$, the coupling constant $g_{\pi\gamma\gamma}$ in (6.04) is determined by the triangle anomaly (the extrapolation to finite m_{π^0} gives a small correction):

$$g_{\pi\gamma\gamma} = \frac{\sqrt{2}\alpha}{\pi f_\pi} N_C <e_q^2>_{\pi^0}. \tag{6.13}$$

The pion decay constant f_π is about 95 MeV and N_C is the number of colour degrees of freedom.

The $\gamma\gamma$ width of the π^0 and the total e^+e^- cross section are the only direct measurements of the number of colour degrees of freedom. Using (6.04) and (6.13) it has been predicted that:

$\Gamma(\pi^0\rightarrow\gamma\gamma) = 7.63$ eV   for $N_C=3$

$\Gamma(\pi^0\rightarrow\gamma\gamma) = 0.85$ eV   for $N_C=1$.

The experimental result /72/

$\Gamma(\pi^0\rightarrow\gamma\gamma) = 7.85 \pm 0.54$ eV

is clearly only consistent with the assumption of 3 colours.

Although the first two-photon experiment proposed for e^+e^- storage rings was the measurement of the π^0 width, this measurement has not yet been done. The reason is that in a storage ring environment the separation of the low energy photons of the π^0 decay from background is difficult even for a specialized detector like the Crystal Ball. The results on

$\Gamma(\pi^0 \to \gamma\gamma)$ have been obtained by lifetime measurements in thin foils and by use of the Primakoff effect.

6.3.2   The Two-Photon Coupling of the η

The measurement of the $\gamma\gamma$ width of the η at e^+e^- storage rings has similar problems as in the case of the π^0. Until recently, only measurements using the Primakoff effect have been available /73, 74/. The Particle Data Group gives as the best value /72/:

$$\Gamma(\eta \to \gamma\gamma) = 0.324 \pm 0.046 \text{ keV}.$$

The error is dominated by systematic uncertainties in separating the coulombic and the nuclear production of the η.

The first measurement of the reaction

$$e^+e^- \to e^+e^-\eta \to e^+e^-\gamma\gamma \qquad (6.14)$$

was reported by the Crystal Ball group /35/. The data, corresponding to an integrated luminosity of 2.7 pb^{-1}, were taken at the SPEAR storage ring at beam energies between 2.5 to 3.5 GeV. The trigger, which was sensitive to the low energy photons of the η decay, required a total energy of more than 475 MeV deposited in two separate clusters with approximate energy balance in the plane transverse to the beam axis. The scattered electrons in (6.14) were not detected (no-tag). A large cosmic ray background could be efficiently reduced by applying timing cuts and by requiring transverse momentum balance. With these cuts the $\gamma\gamma$ mass spectrum in Fig.6.5a was obtained. The remaining background is mainly due to electron scattering off the residual gas in the beam pipe and was determined by running with separated beams. The $\gamma\gamma$ mass spectrum obtained in this way is shown in Fig.6.5b. This spectrum corresponds to roughly 1/3 of the colliding beam data. The indication of an η signal in the separated beam data is consistent with the expected rate from electroproduction.

After background subtraction 56±12 events with a $\gamma\gamma$ mass distribution as shown in Fig.6.5c remain. Practically all of them are compatible with stemming from reaction (6.14). From these events the $\gamma\gamma$ width of the η was derived to be (with statistical and systematic errors):

$$\Gamma(\eta \to \gamma\gamma) = 0.56 \pm 0.12 \pm 0.10 \text{ keV}.$$

It is very important to have the $\gamma\gamma$ width of the η determined with the two different methods because both have their specific systematic problems. However, the error of the storage ring result is still large. A higher statistics measurement by an e^+e^- experiment is necessary to see if the two methods differ in their results.

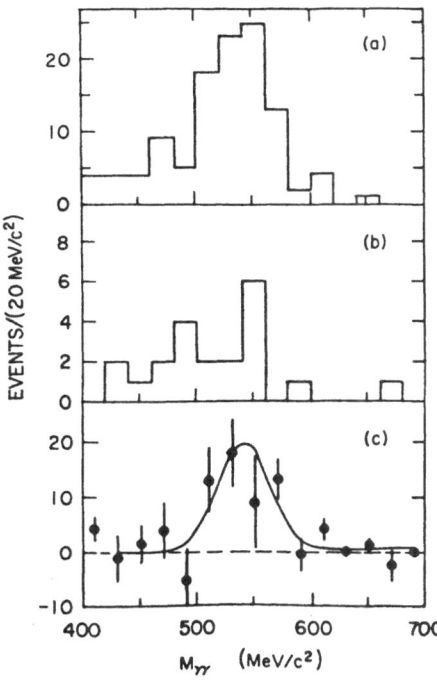

<u>Fig.6.5.</u>   The $\gamma\gamma$ mass distribution for events with two photons detected (Crystal Ball):
a) Colliding beam data.
b) Separated beam data. The scale of the ordinate is adjusted so that this plot is comparable to plot a).
c) Background subtracted data. The curve represents a fit with a Gaussian resolution curve and a small amount of residual background.

6.3.3   The Two-Photon Coupling of the η'

The two–photon production of a meson in an e^+e^- storage ring was first observed at SPEAR by the Mark II group, who measured the $\gamma\gamma$ width of the η' /62/ (upper limits had been obtained before /20, 75/). Subsequently the observation of the η' has also been reported by the JADE /77/ and CELLO /78/ groups at PETRA. All experiments looked for the following decay chain:

$$e^+e^- \to e^+e^-\eta'$$
$$\quad\quad \overset{\llcorner}{\to}\gamma\rho^\circ \to \gamma\pi^+\pi^-. \tag{6.15}$$

The branching ratio of $\eta'\to\gamma\rho^\circ$ is 30% /72/. The Dalitz plot density for the η' decay in (6.15) can be written as a product of the phase space density for the 3 particle final state and the square of a matrix element M:

$$\frac{d^2\sigma}{dM_{\pi\pi}{}^2\ dM_{\pi\gamma}{}^2} = \frac{d^2R^3}{dM_{\pi\pi}{}^2\ dM_{\pi\gamma}{}^2} \cdot |M|^2. \tag{6.16}$$

The matrix element M describes the magnetic dipole transition from a $J^P=0^-$ state to a 1^- state and the formation of the ρ resonance. In the ρ rest system $|M|^2$ is given by /79/:

$$|M|^2 \sim k^2\ q^2\ M_{\pi\pi}{}^2\ \sin^2\vartheta\ [(M_{\pi\pi}{}^2 - M_\rho{}^2)^2 + M_\rho{}^2\ \Gamma^2)]^{-1}; \tag{6.17}$$

k is the photon energy, q the pion momentum and ϑ the angle between a pion and the photon direction (all given in the ρ center of mass system). In /79/ the following parametrization of the ρ width Γ is used:

$$\Gamma = \Gamma_0 \ (q/q_0)^3 \ M_\rho/M_{\pi\pi}. \tag{6.18}$$

With this matrix element the peak of the $\pi\pi$ mass distribution is shifted downward by approximately 25 MeV and the low mass tail of the Breit-Wigner curve is enhanced. Correspondingly, the photon energies are on average larger than expected from a phase space behaviour of the $\rho\gamma$ final state. For most experiments the detection efficiencies are in this case higher. Unfortunately, this had not been realized by the first experiments /62, 77, 78/, which determined the efficiencies with a $\rho\gamma$ phase space.

The TASSO /80/ and the PLUTO group /81/ presented new analyses including the correct matrix element (6.17). The values obtained for the $\gamma\gamma$ width are consistently lower than those of previous analyses. However, these results are still preliminary and one has to wait for confirmation. The average of the published values of the Mark II, JADE and CELLO groups is:

$$\Gamma(\eta' \rightarrow \gamma\gamma) = 5.5 \pm 0.7 \text{ keV} \quad \text{(published)}.$$

The preliminary results from the TASSO and PLUTO groups give an average value:

$$\Gamma(\eta' \rightarrow \gamma\gamma) = 3.8 \pm 0.6 \text{ keV} \quad \text{(preliminary)}.$$

As an example of the experimental method, we explain in the following the measurement of the JADE group. The data used in the analysis correspond to an integrated luminosity of 36 pb^{-1} at beam energies around 17 GeV. Events with two oppositely charged tracks and one photon with an energy larger than 130 MeV were selected. The background from the radiative QED process

$$e^+e^- \rightarrow e^+e^-e^+e^-\gamma \tag{6.19}$$

was reduced by separating pions from electrons by means of dE/dx measurements in the drift chamber (p<0.8 GeV) and shower recognition in the lead glass (p>0.8 GeV). Background from other channels was rejected by demanding the total transverse momentum of the $\pi^+\pi^-\gamma$ system to be less than 600 MeV. The latter cut is rather loose; it does not restrict this measurement to quasi-real photons.

The resulting $\pi^+\pi^-$ and $\pi^+\pi^-\gamma$ mass spectra are shown in Fig.6.6. In the $\pi^+\pi^-\gamma$ mass plot the η' signal stands out clearly. The background is particularly small if the $\pi^+\pi^-$ mass is restricted to the ρ region. The number of η' events was determined by fitting to this distribution the sum of a resolution curve for the η' and a polynomial background curve.

The experimental results for the $\gamma\gamma$ width of the η' are summarized in Table 6.1. Included is the result of a πp scattering experiment which com-

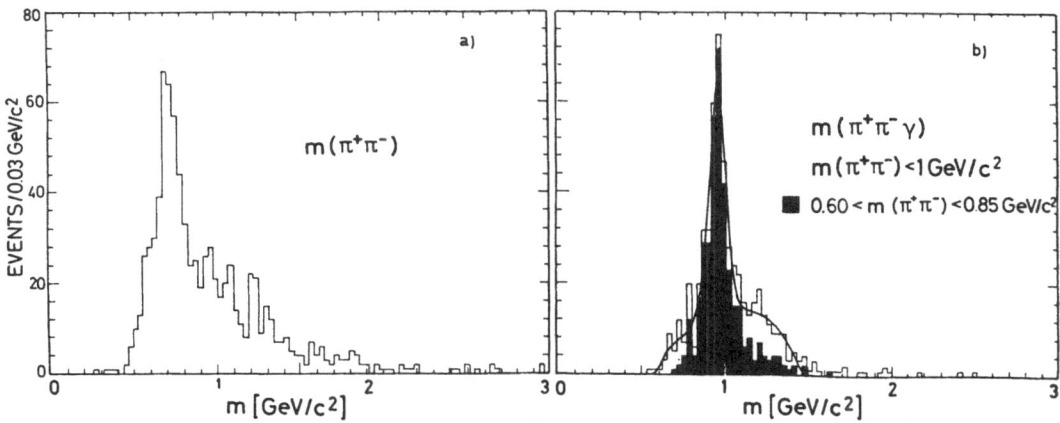

<u>Fig.6.6.</u>   JADE result for $\eta' \to \pi^+\pi^-\gamma$: a) $\pi^+\pi^-$ mass spectrum, b) $\pi^+\pi^-\gamma$ mass spectrum (the curve represents the best fit to the η' peak and the background).

bined a measurement of the total width with the known $\gamma\gamma$ branching fraction to derive the $\gamma\gamma$ width of the η' /76/.

6.3.4   The Q^2 Dependence of the $\gamma\gamma$ Coupling of Pseudoscalars

The two photons in the decay of a pseudoscalar have to be in an anti-symmetric spin state, i.e. the total spin has to be 1. This follows from Bose symmetry of the two-photon wave function and from the negative parity of the pseudoscalars. A total spin 1 can only be composed of transverse photons with the helicity combination (++) or (--). The helicity combination (0,0) of scalar photons does not contribute even for virtual photons. Therefore, the $\gamma\gamma$ vertex function of pseudoscalars, $A^P_{\mu\nu}$ as defined in (2.09), can be written in terms of only one form factor /14/:

$$A^P_{\mu\nu}(Q_1{}^2,Q_2{}^2) = \varepsilon_{\mu\nu\rho\sigma}\, q_1{}^\rho\, q_2{}^\sigma\, F_P(Q_1{}^2,Q_2{}^2). \tag{6.20}$$

For the π^0 the normalization at $Q_1{}^2=Q_2{}^2=0$ is given by the triangle anomaly (6.13):

$$F_{\pi^\cdot}(0,0) = g_{\pi\gamma\gamma}. \tag{6.21}$$

Experimentally, the presence of only one form factor facilitates the analysis, because only one cross section term, σ_{TT}, in (2.14) has to be measured:

$$\sigma_{TT}(W_{\gamma\gamma},Q_1{}^2,Q_2{}^2) = \frac{\sqrt{(q_2{}^2-q_1{}^2)^2-W_{\gamma\gamma}{}^2(q_2-q_1)^2}}{8\sqrt{(q_1q_2)^2-q_1{}^2q_2{}^2}} \;\; \frac{W_{\gamma\gamma}\cdot\Gamma\cdot F_P{}^2(Q_1{}^2,Q_2{}^2)}{(W_{\gamma\gamma}{}^2-M_R{}^2)^2 + \Gamma^2 M_R{}^2}. \tag{6.22}$$

First results on the Q^2 dependence of the η' production were obtained by the PLUTO group /81, 82/. The measurement was done under the single-tag condition, varying the mass of one of the virtual photons while keeping the other small. The measured form factor $F_{\eta'}{}^2(Q_1{}^2, Q_2{}^2 \approx 0)$ is consistent with a ρ pole behaviour in the range $0 < Q_1{}^2 < 1.0$ GeV^2.

6.3.5 Discussion of the Results for Pseudoscalar Mesons

As discussed in Sect.6.3.1, the $\gamma\gamma$ coupling of the π° can be rigorously calculated and is given by the triangle anomaly according to (6.13). Assuming SU(3) nonet symmetry, i.e. equality of the pseudoscalar decay constants, $f_\pi = f_{\eta 8} = f_{\eta 1}$, the $\gamma\gamma$ width of the π° can be used to predict the $\gamma\gamma$ width of the η and η' as a function of the pseudoscalar mixing angle. The triangle anomaly is understood to be dominated by large momenta or small distances /83/. If the same short distance mechanism is responsible for the couplings of all pseudoscalars, no additional mass scale will enter into the couplings. Then the ratios of the coupling constants, or equivalently of $\Gamma_{P\gamma\gamma}/m_P{}^3$, are given, according to (6.12), solely by the effective squared charges (6.10) and the nonet mixing angle Θ:

$$
\frac{\Gamma(\pi^\circ \to \gamma\gamma)}{m_{\pi^\circ}{}^3} : \frac{\Gamma(\eta \to \gamma\gamma)}{m_\eta{}^3} : \frac{\Gamma(\eta' \to \gamma\gamma)}{m_{\eta'}{}^3}
$$
$$
= 3 : (\cos\Theta - \sin\Theta\, 2\sqrt{2})^2 : (\sin\Theta + \cos\Theta\, 2\sqrt{2})^2 \tag{6.23}
$$
$$
= 3 : 1.9 \pm 0.2 : 5.8 \pm 0.8 \quad \text{(exp.)}.
$$

The experimental ratios were obtained using the average $\gamma\gamma$ widths in Table 6.2 (average of published values). Note that with the large mass differences in the pseudoscalar nonet, the $1/m^3$ factors in (6.23) imply an extrapolation by a factor of ~360 between the π° and the η'. Thus the fact that these ratios are of the order of 1 could be seen as an indication that the assumptions made are not unreasonable.

From (6.23) the mixing angle can be determined by relating either the $\gamma\gamma$ width of the η or of the η' to the $\gamma\gamma$ width of the π°:

$$\Theta = -8.2 \pm 2.9° \qquad \text{from } \Gamma(\eta \to \gamma\gamma)$$
$$\Theta = -16.9 \pm 5.6° \qquad \text{from } \Gamma(\eta' \to \gamma\gamma).$$

Within the errors these values are consistent with one another. Note, however, that the preliminary results for the η' (Table 6.1), which have been obtained using the correct matrix element, yield for the mixing angle:

$$\Theta = -28.5 \pm 4.4° \qquad \text{from } \Gamma(\eta' \to \gamma\gamma) \text{ (preliminary)}.$$

This value is no longer consistent with the angle determined from the $\gamma\gamma$ width of the η. To proceed further in this analysis, the experimental uncertainties have first to be settled. However, once the experimental results are improved, we will probably be confronted with the question whether the mass extrapolation in (6.23) is not too much stressed. A clarifying word from theorists would be helpful.

Despite the existing systematic uncertainties one sees clearly that the pseudoscalar singlet-octet mixing is far from being ideal ($\Theta_{ideal}=35.3°$). That is in contrast to the almost ideal mixing observed in the vector and tensor meson nonets (e.g. ω-φ mixing: $\Theta = 38.6\pm0.4°$ /72/). From the quadratic Gell-Mann/Okubo mass formula (GMO),

$$\tan^2\Theta = \frac{4K - \pi - 3\eta}{3\eta' - 4K + \pi},$$ (6.24)

(the particle symbols stand for the squared masses of these particles) one calculates for the mixing angle /72/:

$$\Theta = -11.1\pm0.2° \qquad \text{(quadratic GMO)}.$$

The GMO mixing angle is in the range of values derived from the $\gamma\gamma$ widths. The application of the GMO to the pseudoscalar nonet is not without problems, because the mass relations between the octet and singlet are completely inconsistent. From the mass formula for the octet η follows that the mass of the physical η (548.8 MeV) is close to the octet mass:

$$\eta_8 = \tfrac{1}{3}(4K - \pi) \approx 600 \text{ MeV}.$$ (6.25)

This is consistent with the mixing angle being small. However, the corresponding expression for the singlet,

$$\eta_1 = \tfrac{1}{3}(2K + \pi) \approx 400 \text{ MeV},$$ (6.26)

yields a mass which is not at all close to the η' mass (957.6 MeV). Even the sign of the mass difference comes out wrong: the η' should be lighter than the η. This difficulty has been known for a long time and is related to the "U(1) problem" /84/. In view of this mass problem it is particularly important to study the quark composition of the pseudoscalars in electromagnetic transitions.

In the foregoing discussion we have tacitly assumed that the quarks have the conventional fractional charges. Can the measurement of the two-photon coupling of the resonances help to distinguish between the fractionally (FCQ) and the integrally charged quark model (ICQ) ?

In quark models with a global $SU(3)_{flavour} \times SU(3)_{colour}$ symmetry, quarks of the same flavour can in general have different charges for different colours. As a consequence the photon can be decomposed into a colour singlet and a colour octet (see e.g. /85/). The colour octet part is absent only in the Gell-Mann/Zweig scheme of fractionally charged quarks. Below the threshold for open colour a single photon can only couple to the colour singlet hadrons via its colour singlet part. But the colour singlet photon 'sees' only the charges averaged over the colours, which turn out to be always the same as the conventional fractional charges. Thus one needs the colour octet photon to measure the quark charges !

In the Han–Nambu model with integral charges of the quarks /86/, which is a special choice of the $SU(3)_{flavour} \times SU(3)_{colour}$ symmetry, the photon has the following decomposition:

$$\gamma_{ICQ} = (8_f, 1_c) - (1_f, 8_c). \tag{6.27}$$

The flavour octet – colour singlet, ($8_f, 1_c$), is the usual representation of the photon, whereas the flavour singlet – colour octet, ($1_f, 8_c$) is specific to the ICQ model. Since below colour threshold the observed hadrons have to be in a colour singlet state, the colour octet component of the photon does not contribute in processes involving only one photon. In such processes the cross sections obtained in the FCQ and in the ICQ model are the same. However, in higher order electromagnetic processes colour octet photons can combine to colour singlets. For example, the product of two colour octets contains a colour singlet according to the decomposition:

$$(1_f, 8_c) \times (1_f, 8_c) = (1_f, 1_c) + \dots . \tag{6.28}$$

The colour singlet in this decomposition is also a flavour singlet. Therefore, only the two-photon couplings to flavour singlets are sensitive to the differences between the ICQ and the FCQ models, whereas the two-photon couplings of the flavour octets probe the quark charges averaged over colours. On the other hand, applying dispersion relations to the $\gamma\gamma$ decay of a meson, one finds that below colour threshold the contributions from colour octet photons should be suppressed by energy denominators in the dispersion integral. However, it has been argued that there is not such a suppression in the case of the pseudoscalars because their $\gamma\gamma$ couplings, as determined by the triangle anomaly, are dominated by large momenta /87/. The $\gamma\gamma$ width of the singlet (η_1) would then be four times larger in the ICQ than in the FCQ model. With the assumption that $SU(3)_{flavour}$ nonet symmetry holds and that the η' is mainly a singlet we calculate from the π° width /85/:

$\Gamma_{\eta'\gamma\gamma}{}^{FCQ} \approx 6$ keV

$\Gamma_{\eta'\gamma\gamma}{}^{ICQ} \approx 24$ keV.

Hence, with the assumptions made, the ICQ model is excluded by the measured $\gamma\gamma$ width of the η'. However, the necessary assumptions are controversial, in particular the SU(3) nonet symmetry (see the discussion in /87, 88, 89/). Before this controversy is settled, the measured $\gamma\gamma$ width of the η' cannot be used as an absolute proof for fractionally charged quarks. On the other hand, the data give no hint for integral charges.

6.3.6   Upper Limits on the Two-Photon Width of the $\iota(1440)$

With the discovery of the $\iota(1440)$ which most likely has spin-parity 0^- /90/, a new aspect came into the discussion about the pseudoscalars. This state is thought to be a glueball candidate, mainly because it has a large production rate in radiative J/ψ decays $(J/\psi \to \gamma \iota)$, which is a 'glueball favoured' channel as explained above.

The classification of the $\iota(1440)$ is widely discussed in the literature (for a recent review see e.g. /91/). Some authors argue that it is just a radial excitation of the η', others explain the ι as a gluonium (or dominantly gluonium) state. In the latter case one would naively expect a small coupling to $\gamma\gamma$ and the size of the $\gamma\gamma$ coupling could be used to decide between gluonium and quarkonium as suggested in /92/. However, not all theorists seem to agree on that. In fact, it has been argued that the ι could have a relatively large $\gamma\gamma$ width (up to 50 keV /93/) due to mixing with the η' and the special situation in the pseudoscalar sector (strong axial anomaly etc.) /94/. According to VMD, a large $\gamma\gamma$ width will be accompanied by a large $\rho^\circ\rho^\circ$ width. In the model of /94/ it was predicted (for $\Gamma_{tot}(\iota)=60$ MeV):

$$\Gamma(\iota \to \gamma\gamma)\cdot B(\iota \to \rho^\circ\rho^\circ) = 8.0 \text{ keV} \qquad \text{(theor.)}.$$

In the analysis of $\gamma\gamma \to \rho^\circ\rho^\circ$, the TASSO collaboration fitted $J^P=0^-$ intensity (Sect.6.7) and found the upper limit (c.l. = confidence level):

$$\Gamma(\iota \to \gamma\gamma)\cdot B(\iota \to \rho^\circ\rho^\circ) < 1.0 \text{ keV} \qquad (95\% \text{ c.l.}).$$

This limit is in contradiction to the prediction of /93, 94/, but for most of the models explaining the $\iota(1440)$ it is not stringent enough.

The $\iota(1440)$ was first observed via the decay into $K\bar{K}\pi$. The Mark II /24/ and TASSO /80/ collaborations searched in this channel for the ι produced in two-photon reactions. They did not observe a signal and obtained the upper limits given in Table 6.2.

6.4   The Tensor Mesons

6.4.1   Introduction

The 2^{++} nonet of mesons is usually referred to as the tensor meson nonet. The angular momentum configuration of the quarks is $^3P_2$, that means the quarks are in a relative P wave with parallel spins. The other two multiplets of the P wave triplet are the scalars (0^{++}) (see Sect.6.5) and the axial vectors (1^{++})[3]. All members of the 2^{++} nonet (f(1270), A_2(1320),

---

[3]   The axial vectors will not be discussed here because they do not couple to two real photons. However, it should be mentioned that a measurement of the coupling of axial vector mesons to off-shell photons would be very interesting and probably also feasible /95/.

f'(1515), K*(1430)) are well established /67/. Here we are interested in the flavour neutral states: the isovector A_2, the isoscalar f with mainly u$\bar{u}$ and d$\bar{d}$ content and the isoscalar f' with mainly s$\bar{s}$ content.

In the case of the tensor mesons there is no absolute prediction for the $\gamma\gamma$ couplings from an anomaly theorem, as in the case of the pseudoscalars. Various theoretical attempts have been made to calculate the $\gamma\gamma$ widths using the quark model, finite energy sum rules or dispersion relations /96 to 104/. A summary of the predictions for the f partial width has been given in /30/; the values lie between about 1 and 20 keV. The predictions for the ratios of the $\gamma\gamma$ couplings within the tensor meson nonet stand on more solid ground than the predictions for the absolute values. With the assumption of SU(3) nonet symmetry and fractionally charged quarks one gets for ideally mixed tensor mesons the ratios:

$$g_{A_2\gamma\gamma}^2 : g_{f\gamma\gamma}^2 : g_{f'\gamma\gamma}^2 = 9 : 25 : 2. \tag{6.29}$$

This relation holds for each $\gamma\gamma$ helicity λ ($|\lambda|=0$, 2 for real photons) separately. Deviations from these ratios would indicate, under the assumptions made above, that the f and the f' are not ideally mixed states.

In the following we discuss how one can relate the $\gamma\gamma$ coupling constants in (6.29) to the measured $\gamma\gamma$ widths. The decay of a tensor meson T into two real photons is determined by two independent amplitudes, which may be taken as the $\lambda=0$ and $\lambda=2$ $\gamma\gamma$ helicity amplitudes. It has been predicted (see discussion below) that the $\gamma\gamma$ helicities $\lambda=\pm2$ (photons with opposite helicity) should dominate over $\lambda=0$ (photons with same helicity). Neclecting $\lambda=0$ contributions, the $\gamma\gamma$ width depends on one coupling constant, $g_{T\gamma\gamma}$, only. In terms of this coupling constant the $\gamma\gamma$ width has been written /105/:

$$\Gamma_{T\gamma\gamma} = \frac{g_{T\gamma\gamma}^2}{4\pi} \frac{m_T^3}{80\mu^2}, \tag{6.30}$$

where μ is assumed to be a common scale for the tensor meson nonet. With this assumption one has:

$$\frac{\Gamma_{T\gamma\gamma}}{m_T^3} \sim g_{T\gamma\gamma}^2. \tag{6.31}$$

We have not seen a rigorous derivation that the SU(3) relation (6.29) applies to $\Gamma_{T\gamma\gamma}/m_T^3$. In /103/ the ratios in (6.29) have been applied to the dimensionless quantity $\Gamma_{T\gamma\gamma}/m_T$ (corresponding to $\mu=m_T$ in (6.30)). For the tensors, the mass dependence is not as dramatic as for pseudoscalars. A cubic mass term changes by a factor ~1.7 between the f and the f', which must be compared to a factor ~360 between the π° and the η'. The fact that this huge extrapolation works for pseudoscalars is usually explained by the short distance, scale invariant character of the triangle anomaly, which is assumed to dominate the pseudoscalar $\gamma\gamma$ couplings.

A prediction, which seems to be rather independent of the different theoretical approaches, is that the tensor meson production by two real photons proceeds predominantly via a $\gamma\gamma$ helicity $\lambda=2$, i.e. via two photons with opposite helicity. The $\lambda=2$ dominance is obtained just from Clebsch-Gordan coefficients if one assumes that only the lowest multipole in the $\gamma\gamma$ system contributes. In this case one finds for the ratios of the $\lambda=0$ and $\lambda=2$ intensities ($\lambda=1$ does not contribute for real photons):

$$I(\lambda=0) : I(\lambda=2) = 1 : 6. \tag{6.32}$$

Helicity 2 dominance has been derived under more general assumptions by several authors, e.g. /97, 103/. In /103/ a sum rule is given relating the cross sections for the helicity $\lambda=0$ and $\lambda=2$ projections of partial waves with $J\geq2$. Assuming narrow resonance saturation one gets the following inequality ($g_{T\gamma\gamma}(\lambda)$ is the coupling constant of tensor mesons for helicity λ):

$$\sum_T \frac{g_{T\gamma\gamma}^2(\lambda=2)}{m_T^8} \geq 6 \sum_T \frac{g_{T\gamma\gamma}^2(\lambda=0)}{m_T^8} + \text{(positive contributions from } J^P=4^+). \tag{6.33}$$

This inequality implies the dominance of the $\gamma\gamma$ helicity $\lambda=2$. None of the experiments so far is in contradiction to this prediction. However, in most cases the experiments have not been sensitive to the different helicity contributions. Therefore the experimenters usually rely in their analyses on the assumption of $\lambda=2$ dominance (an exception is the Crystal Ball analysis of $\gamma\gamma\to f$). The angular distribution of a tensor meson in a helicity λ state decaying into two (pseudo)scalar mesons is $\sim|Y_2^\lambda(\cos\vartheta^*)|^2$ (ϑ^* is the polar angle of one of the decay particles with respect to the $\gamma\gamma$ direction in the $\gamma\gamma$ center of mass system):

$$\lambda = 0 : \quad |Y_2^0(\cos\vartheta^*)|^2 \sim (\cos^2\vartheta^* - \tfrac{1}{3})^2$$
$$\lambda = 1 : \quad |Y_2^1(\cos\vartheta^*)|^2 \sim \sin^2\vartheta^* \cos^2\vartheta^* \tag{6.34}$$
$$\lambda = 2 : \quad |Y_2^2(\cos\vartheta^*)|^2 \sim \sin^4\vartheta^*.$$

The angular distributions for $\lambda=0$ and $\lambda=2$ are experimentally hard to distinguish because they have the same shape around 90°, where the detectors have the best acceptance (see Fig.6.7).

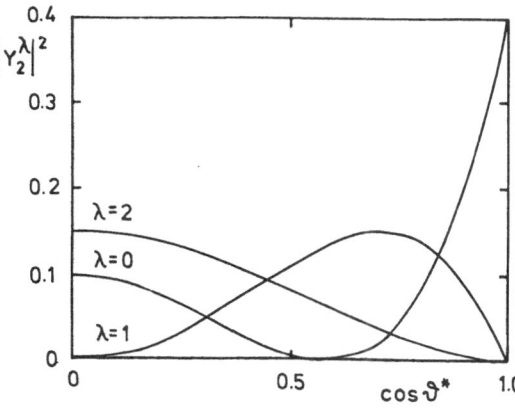

Fig.6.7. Decay angular distributions of tensor mesons decaying into two (pseudo)scalar mesons for different $\gamma\gamma$ helicities λ.

For the case of f production with one of the photons off-shell, the authors of /106/ present a model for the evolution of the different helicity contributions with increasing Q^2 of the off-shell photon. The minimum tagging angle plotted in Fig.6.8 is related to the experimentally accepted Q^2 range. It is interesting to note that in the model of /106/ the helicities $\lambda=0$ and especially $\lambda=1$ are suppressed only at very low Q^2. As predicted in /14/, helicity $\lambda=0$ should even dominate over $\lambda=2$ at large Q^2.

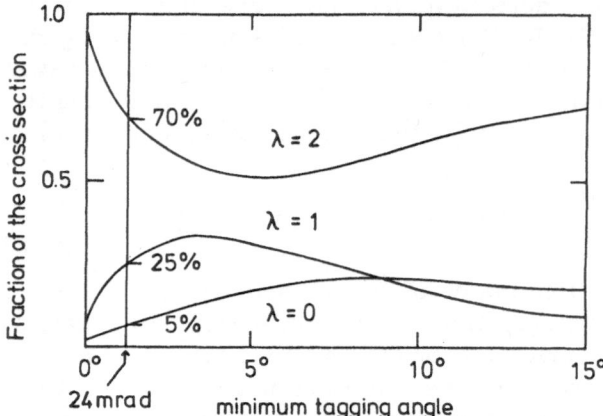

Fig.6.8. Predicted helicity 0,1,2 contributions to $\gamma\gamma\to f$ as a function of the minimum tagging angle (0° corresponds to the no-tag condition).

6.4.2 The Two-Photon Coupling of the f(1270)

The f(1270) is the isoscalar in the 2^{++} nonet with mainly u and d quark content, as can be inferred from the dominance of the two pion decay mode (branching ratio: 83.1±1.9% /72/). The experiments PLUTO, TASSO, Mark II and CELLO observed the charged decay mode $f\to\pi^+\pi^-$ and Crystal Ball and JADE the neutral mode $f\to\pi^0\pi^0$.

For the two-photon production of two charged pions (as well as other charged particle pairs) one expects a large non-resonant continuum arising from the direct coupling of the photons to the charges. Close to threshold continuum production may be sufficiently well described by the Born approximation as given by the three diagrams in Fig.5.1. In Chap.5 we showed that the Born cross section for $\gamma\gamma\to\pi^+\pi^-$ is dominated in the f region by the $\lambda=2$ helicity contribution (Fig.5.3). We also argued, that up to the f region the helicity 2 amplitude may be well described by the Born approximation because the $\pi\pi$ phase shifts for $J\geq2$ are small below the f. Hence we may approximate $\pi^+\pi^-$ production in the f region by adding to the Born diagrams in Fig.5.1 a resonance contribution (Fig.6.9). Since the f is also predicted to be dominantly produced in a $\lambda=2$ state, one may ex-

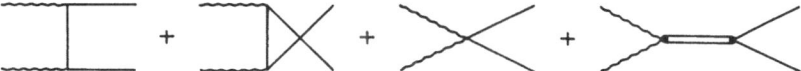

Fig.6.9.   Basic diagrams for $\gamma\gamma \to \pi\pi$ including the $\gamma\gamma$ coupling to a resonance.

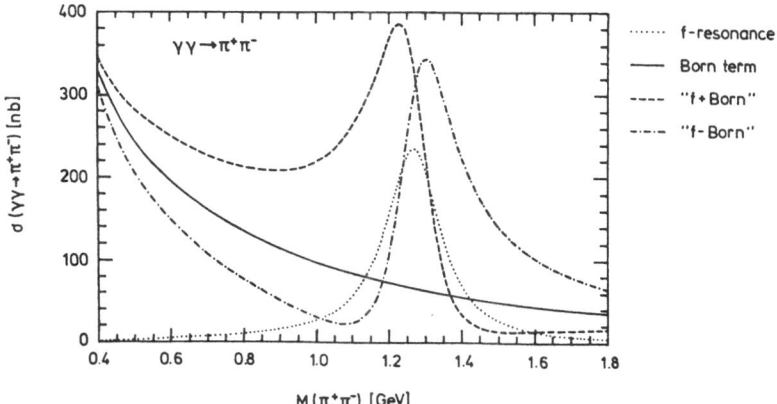

Fig.6.10.   Cross sections for two-photon production of $\pi^+\pi^-$ according to the Born approximation and to f resonance production : The f resonance (dotted curve) and the Born term (full curve) are given separately as well as coherently added (dashed curve: with the same phase, dashed-dotted curve: with opposite phase).

pect a strong interplay between the f resonance and the $\pi^+\pi^-$ continuum[4]. The interference is completely determined by the Breit-Wigner phase (the Born amplitude is purely real) except for a relative sign between the resonance and the Born amplitude.

The cross sections obtained in the f region from this model are shown in Fig.6.10 for three different cases ($\gamma\gamma$ width of the f is taken from Table 6.2): a) the resonance contribution alone, b) constructive and c) destructive interference of the resonance with the Born term. Here 'constructive' and 'destructive' refer to the relative sign between the resonance and Born graphs. The angular distribution arising from constructive interference (which is favoured by the data, see below) is plotted in Fig.6.11 for different $W_{\gamma\gamma}$ values in the f region.

Many authors constructed models to improve this simple approach for $\gamma\gamma$ production of pion and kaon pairs up to invariant masses of about 1.5 GeV, see e.g. /49, 50/. The CELLO group used for their analysis of the f

---

[4]   If the lepton scattering planes are not measured, interference is observable only for the same $\gamma\gamma$ helicity states. As a consequence, interferences between states with the same J can only be seen in the total cross section and not in the angular distributions.

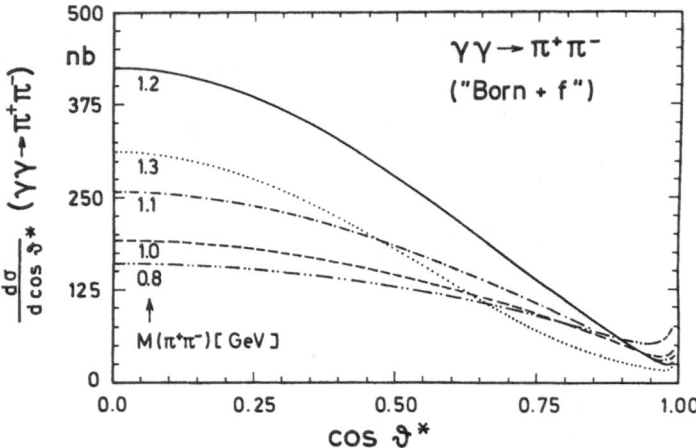

<u>Fig.6.11.</u>   Angular distribution of the pions from f decay (λ=2) assuming constructive interference with the Born contribution.

resonance (see below) a program developed by Mennessier /50/. In this program the Born diagrams and the direct coupling to resonances are unitarized by considering the measured phase shifts in the coupled channels $\pi^+\pi^-$, $\pi^0\pi^0$, $K\overline{K}$ (Fig.5.4). In addition to the π exchange diagrams, ρ and ω exchange diagrams were also included. With this approach, one gets in principle also predictions for the $\pi^0\pi^0$ and $K\overline{K}$ channels. However, the $\pi^+\pi^-$ channel, dominated by the Born term and the f resonance, is probably described more reliably. Especially $K\overline{K}$ production is rather complicated due to possible resonant threshold enhancements and isovector contributions (e.g. δ(980), A_2(1320)).

<u>Measurements of f$\rightarrow\pi^+\pi^-$:</u>
The analysis of two-photon production of the f via the two charged pion decay mode is complicated by large backgrounds from $\gamma\gamma$ QED processes and from the $\pi^+\pi^-$ continuum. Candidate events for f production are selected from events with two oppositely charged tracks. Such a sample contains also events with electron and muon pairs from the two-photon QED reactions $e^+e^-\rightarrow e^+e^-e^+e^-$ and $e^+e^-\rightarrow e^+e^-\mu^+\mu^-$. For most of the detectors it is difficult or impossible to separate low energy (< 1 GeV) electrons and muons from pions. The QED processes can be reliably calculated (up to radiative corrections, which are estimated to be smaller than 10% /48/) and subtracted from the data. The Monte Carlo program written by Vermaseren /38/ is commonly used to calculate the QED background.
The first measurements of the $\gamma\gamma$ width of the f meson were done by PLUTO /43/ and then by TASSO /44/. In both experiments the mass spectra of events with 2 charged tracks show in the f region a clear enhancement above the QED background. The mass spectrum in Fig.6.12 was obtained by TASSO assuming pion masses for the two particles. After subtraction of the QED background, the distribution of the pion production

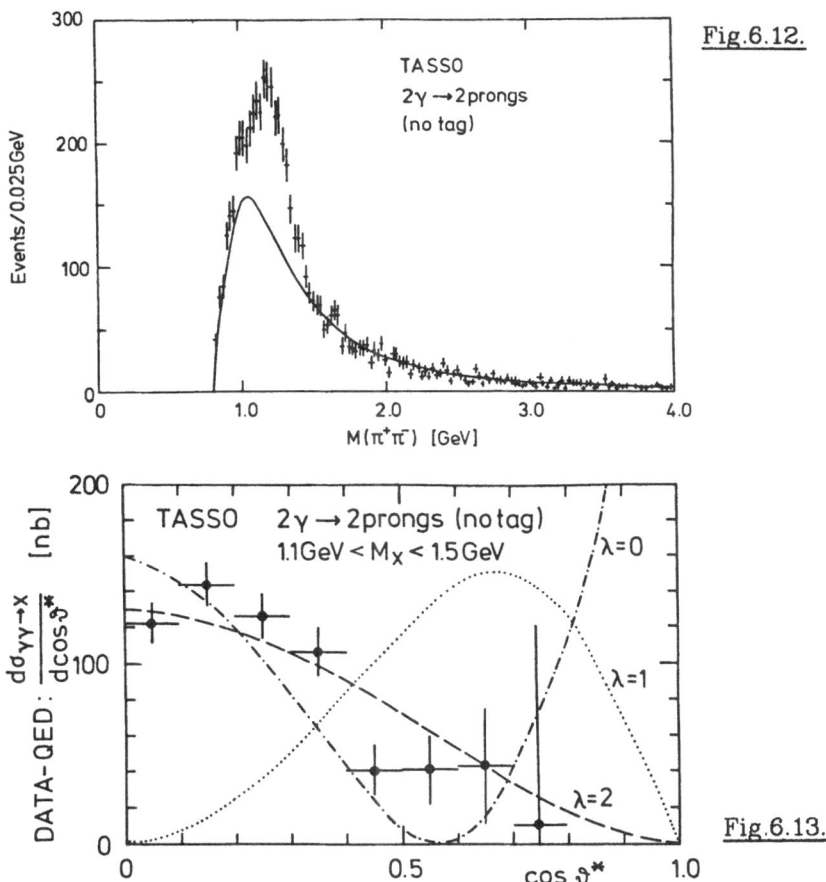

Fig.6.12.

Fig.6.13.

Fig.6.12.   Invariant mass distribution of events with two charged tracks obtained by assigning pion masses to all tracks (TASSO). The curve is the QED prediction for lepton pair production by two photons.

Fig.6.13.   Differential cross section for $\gamma\gamma \to \pi^+\pi^-$ in the f region measured by TASSO. The curves show the expectations for the different $\gamma\gamma$ helicities.

angle is consistent with a $\sin^4\vartheta^*$ behaviour as expected if helicity 2 dominates (Fig.6.13). Unfortunately, in these first experiments the $\pi^+\pi^-$ mass spectra are cut off at about 1 GeV by the transverse momentum threshold of the triggers. A detailed study of the $\pi^+\pi^-$ continuum below the f was therefore not possible. In both measurements there is little room for a continuum above the f (see also Fig.4.8 in Chap.4). The TASSO group finds an excess above the QED background below the f, which cannot be assigned to the f, and an apparent mass shift of the f of the order of 30 MeV, suggesting an interference with the continuum. However, it was impossible to decide upon the presence of a continuum because neither a simple Born term, which was too high above the f region, nor a Born term with some type of absorption correction was able to describe the data. The TASSO group fitted their data with a f resonance using standard values for mass and width and in addition a S*(975) resonance to account for the additional $\pi^+\pi^-$ contribution. This procedure may have effectively accounted for the continuum background, although the size of the S* contribution required by the fit has been ruled out by the Crystal Ball data (see below).

The TASSO collaboration also studied the Q^2 dependence of the $f\gamma\gamma$ coupling /44/. Using single-tag data, with an average $Q^2 \approx 0.35$ GeV² for the tagged photon, the $\pi^+\pi^-$ mass spectrum in Fig.6.14 was obtained. Because of the expected complicated helicity structure of f production and because of the limited angular acceptance, the $\gamma\gamma$ coupling could not be analysed in a model independent way. The TASSO group used a nonrelativistic quark model /106/ which describes the Q^2 dependence of the helicity contributions for $\lambda=0$, 1, 2, the only free parameter being the $\gamma\gamma$ width for real photons. For one virtual photon with $Q^2 \approx 0.35$ GeV² the model predicts a suppression of f production by a factor of about 0.5 relative to $Q^2 = 0$. The single-tag data require the $\gamma\gamma$ width for real photons (i.e. extrapolated to $Q^2=0$) to be:

$$\Gamma(f\to\gamma\gamma) = 1.6 \pm 0.6 \pm 0.3 \text{ keV},$$

which has to be compared to the no-tag result (Table 6.1):

$$\Gamma(f\to\gamma\gamma) = 3.2 \pm 0.2 \pm 0.6 \text{ keV}.$$

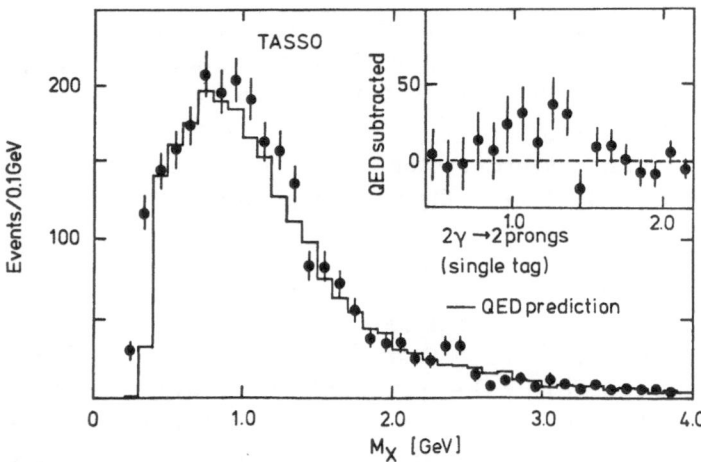

Fig.6.14.   Single-tag   measurement   of $e^+e^-\to e^+e^-$+two   charged particles (TASSO). The plot shows the distribution of the invariant mass of the two particles detected in the central detector assuming pion masses. The insert shows the same distribution with the QED contribution subtracted.

The confidence level for both widths being the same is 8%. Higher statistics and a better angular acceptance are needed to test the prediction of the model for the Q^2 dependence of the helicity structure of f production.

The Mark II experiment at SPEAR has published a f analysis /45/ with a $\pi^+\pi^-$ mass spectrum extending down to 0.5 GeV, thus allowing the analysis of the continuum below the f. This work has an interesting experimental feature: The rate of produced pion pairs was determined directly by identifying pions. The identification method exploited the fact that pions have a relatively high probability to be absorbed in the magnet coil in front of the liquid argon calorimeter. In Fig.6.15 the cross section $\sigma(\gamma\gamma\rightarrow\pi^+\pi^-)$ is shown for $|\cos\vartheta^*|<0.35$ where ϑ^* is the center of mass polar angle of the π's. The data are compared to the Born term and to a QCD prediction /107/ for the two-photon production of pion pairs at large angles and high energies. The excess of events above the Born cross section can be described by a Breit-Wigner curve for the f resonance. At the high mass side this experiment is not as sensitive as the PETRA/PEP experiments because of the limited $\gamma\gamma$ flux (beam energy ~3 GeV).

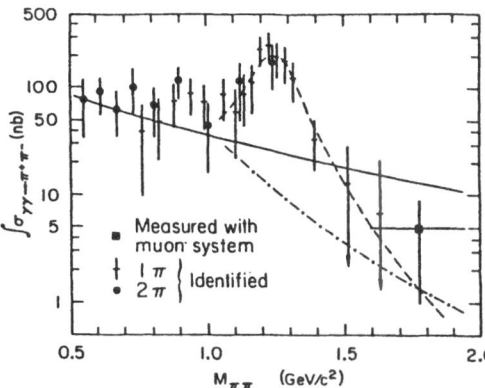

<u>Fig.6.15.</u>   Cross section $\sigma(\gamma\gamma\rightarrow\pi^+\pi^-)$ for $|\cos\vartheta^*| < 0.35$ from events with identified pions (Mark II). The data are compared to the Born approximation (solid curve), a Breit-Wigner curve for the f (dashed curve) and a QCD calculation /107/ (dashed-dotted curve).

The $\gamma\gamma$ width of the f was derived from the mass spectrum of the full 2-prong sample. The identified pions were only used to determine the absolute normalization of the QED background. The subtracted mass spectrum (Fig.6.16) was fitted by a sum of interfering contributions from a f Breit-Wigner amplitude with standard resonance parameters and the Born term. An extra term was added to account for misidentified K's from f'(1515)$\rightarrow$K$^+$K$^-$. In the notation of /45/ the total $\pi^+\pi^-$ cross section as a function of the mass reads:

$$\pi\pi(m) = f(m) + \alpha \cdot C(m) + 2\beta \cdot \cos\delta(m) \cdot \sqrt{\alpha \cdot C(m) \cdot f(m)} + f'(m). \qquad (6.35)$$

The symbols $\pi\pi$, f, f', C stand for the total cross section for $\pi^+\pi^-$ and for the f, f' and Born term contributions, respectively. Since the Born term is real, the interference is determined by the phase shift $\delta(m)$ of the f resonance (modulo π). The parameter β ($-1<\beta<1$) was included because no par-

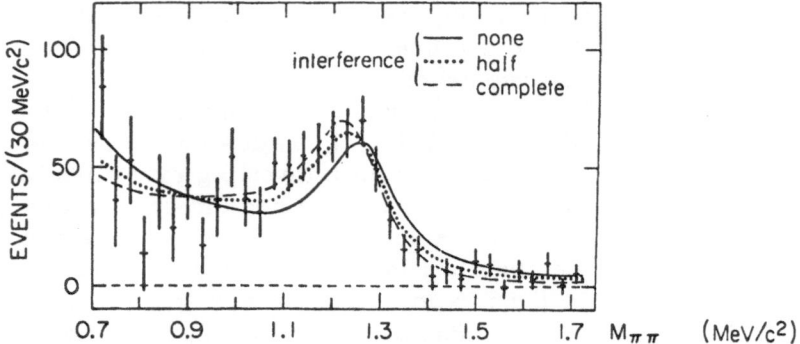

<u>Fig.6.16.</u>   Invariant mass distribution of events with two charged tracks obtained by assigning   pion masses to all tracks (Mark II). The QED background is subtracted.   The data are compared to an interfering sum of a Breit-Wigner   and the $\pi^+\pi^-$ Born term. The different curves are explained in the text.

tial wave decomposition had been done, so as to determine the interfering amplitudes. Data required β to be positive (constructive interference below the f peak, destructive above). Examples for fits with different values of β (β=0, 0.5, 1.0) are shown in Fig.6.16. The resulting $\gamma\gamma$ width of the f (see Table 6.1) does not depend sensitively on the value of β. A fit with full interference gives a good description of the data, i.e. it reproduces the observed resonance shape and the mass shift of about 30 to 40 MeV. Full interference is in accordance with the dominance of the (J=2, λ=2) partial wave of the Born term (Fig.5.3), if the f resonance is also dominated by λ=2.

The CELLO results on $\pi^+\pi^-$ production between 0.8 and 1.5 GeV /80, 81/ have been described by the model of Mennessier /50/. The solid curve in Fig.6.17 has been calculated including only the unitarized Born term and

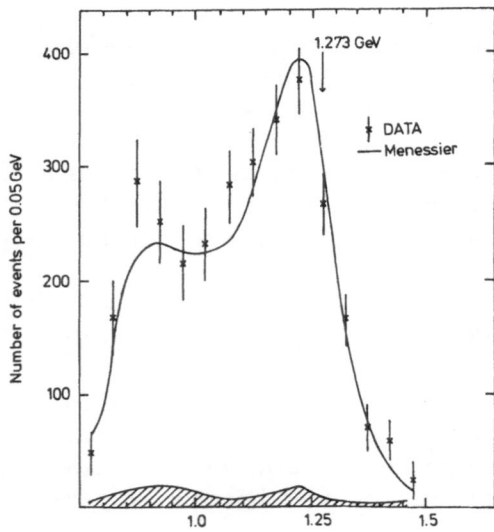

<u>Fig.6.17.</u>   Invariant mass distribution of events with two charged tracks obtained by assigning pion masses to all tracks (CELLO). The QED background is subtracted. The hatched area is the estimated K^+K^- background. The solid curve presents a Monte Carlo simulation using the model of Mennessier /50/ with $\Gamma(f\to\gamma\gamma)$ = 2.7 keV (see text).

the coupling to the f resonance from Fig.5.4. In particular, a contribution of a scalar resonance was not required by the data. The modification of the Born term obtained from the unitarization procedure using strong interaction data turns out to be small. This may be due to the smallness of the D wave phase shifts below the f mass. The model explains the observed shift of the f peak by about 40 MeV in the $\pi^+\pi^-$ channel and predicts also a small downward shift in the $\pi^0\pi^0$ channel which is consistent with the observation of the Crystal Ball group (see next section).

Measurements of $f\to\pi^0\pi^0$:

Measuring the two-photon production of the f via the $\pi^0\pi^0$ final state has the advantage that there is no background from QED reactions. Continuum production of $\pi^0\pi^0$ should also be much smaller than in the charged case because direct coupling of the photons to the neutral pions is not possible (in the sense of the Born approximation used in the charged pion case).

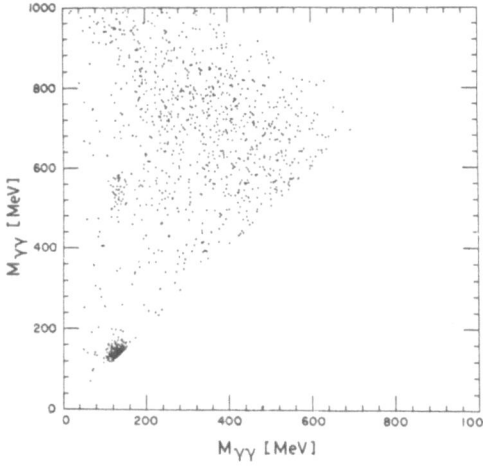

Fig.6.18. Measurement of $\gamma\gamma\to\pi^0\pi^0$ and $\gamma\gamma\to\pi^0\eta$ (Crystal Ball): Scatter plot of the larger $\gamma\gamma$ mass versus the smaller in events with four detected photons.

Results on $\gamma\gamma\to\pi^0\pi^0$ have been published by the Crystal Ball /108/. For events with 4 detected photons, Fig.6.18 shows a scatter plot of one photon pair mass versus the other. The signal in the $\pi^0\pi^0$ region is clearly visible. The enhancement in the $\pi^0\eta$ region will be discussed below in context with the A_2 meson. The $\pi^0\pi^0$ mass distribution (Fig.6.19a) exhibits a f signal with little background. A comparison with a Monte Carlo simulation of the reaction $\gamma\gamma\to f\to\pi^0\pi^0$ with standard f resonance parameters indicates that the peak in the data is shifted towards smaller masses by about 35 MeV (solid curve in Fig.6.19b). A fit with the mass and width of the f as free parameters gives M=1238±14 MeV and Γ=248±38 MeV (dashed curve). Using the Mennessier program /50/, with the parameters determined by the CELLO group for the $\pi^+\pi^-$ data, a mass shift in the $\pi^0\pi^0$ channel of about 20 MeV, consistent with the observation, is predicted /80/. On the other hand, the

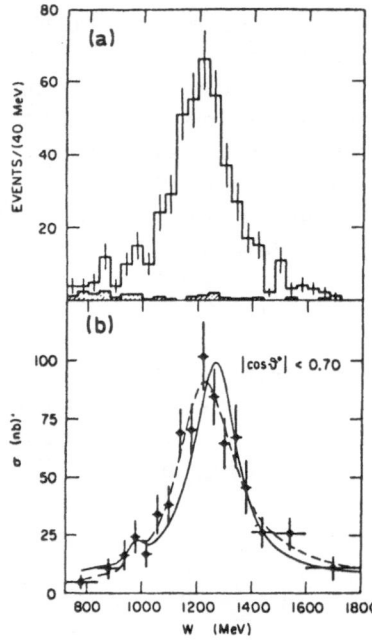

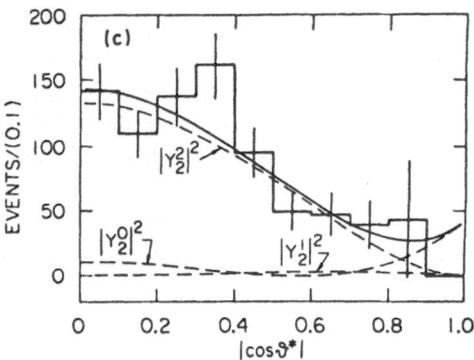

<u>Fig.6.19.</u>   Measurement of $\gamma\gamma\to\pi^\circ\pi^\circ$ by the Crystal Ball: a) $\pi^\circ\pi^\circ$ mass spectrum, b) $\gamma\gamma\to\pi^\circ\pi^\circ$ cross section and c) decay angular distribution of $f\to\pi^\circ\pi^\circ$.

estimated systematic error in the mass measurement of about 2%, together-er with the statistical error, could also account for the observed mass shift. The fit also allowed for an S*(975) contribution. No evidence for such a signal was found. The upper limit derived is given in Table 6.2 (see also Sect.6.5).

The Crystal Ball group also measured the helicity structure of f pro-duction. The large angular acceptance allowed the determination of the helicity contributions by fitting to the $\cos\vartheta^*$ distribution (Fig.6.19c) the expression:

$$f(|\cos\vartheta^*|) = a_0\cdot|Y_2^0|^2 + a_1\cdot|Y_2^1|^2 + a_2\cdot|Y_2^2|^2. \tag{6.36}$$

The result,

$a_0/a_2 = 0.12\pm0.39,$

$a_1/a_2 = 0.02\pm0.11,$

is in agreement with helicity 2 dominance. The $\gamma\gamma$ width of the f deter-mined with the fitted helicity structure,

$\Gamma(f\to\gamma\gamma) = 2.9 \pm 0.6 \pm 0.6$ keV   (helicity fitted)

is consistent with the value given in Table 6.1, which was obtained with the assumption that only helicity 2 contributes.

The JADE collaboration presented preliminary results from an analysis of $f\to\pi^\circ\pi^\circ$ /80/. The plot in Fig.6.20a of one $\gamma\gamma$ mass versus the other in events with 4 γ's detected shows a strong signal of correlated $\pi^\circ\pi^\circ$ pro-duction. The $\pi^\circ\pi^\circ$ mass distribution is shown in Fig.6.20b. Clearly, there is

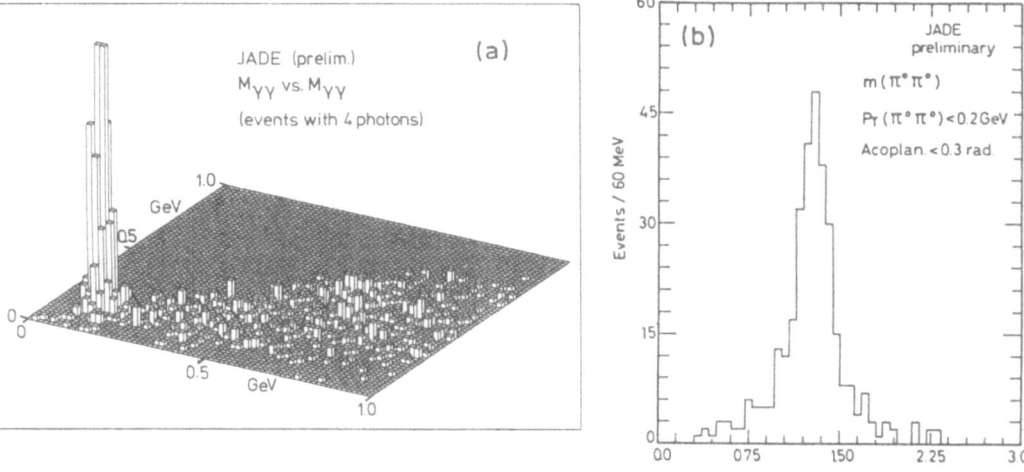

Fig.6.20. Measurement of $f \to \pi^0 \pi^0$ by JADE: a) Scatter plot of one $\gamma\gamma$ mass versus the other in events with 4 photons detected. b) $\pi^0 \pi^0$ invariant mass distribution obtained from a) by requiring both $\gamma\gamma$ masses to lie in the π^0 band (90-190 MeV).

no large continuum visible below the f mass as in the charged pion channel, although the detection efficiency is reasonably large down to about 0.5 GeV. The $\pi^0 \pi^0$ channel is particularly suited to search for the scalar resonances ε and S^*. Upper limits for narrow resonances have been given /80/ (see Sect.6.5).

Summary of the results on f(1270) production:

The production of the f meson by two photons has been observed in the two decay channels $f \to \pi^+ \pi^-$ and $f \to \pi^0 \pi^0$. The charged pion final state is characterized by a strong interference between the continuum and the resonance leading to a downward shift of the resonance peak by about 40 MeV. The mass spectrum and the angular distributions can be explained by a coherent sum of the Born approximation for $\gamma\gamma \to \pi^+ \pi^-$ and a Breit-Wigner amplitude for the f. Unitarization of the amplitudes according to the Mennessier model does not change the $\pi^+ \pi^-$ production cross section by much. The observed mass shifts of the resonance peak both in the $\pi^+ \pi^-$ and the $\pi^0 \pi^0$ channel are in agreement with the prediction from this model. All observed angular distributions of the pions are consistent with helicity 2 dominance. The listed $\gamma\gamma$ widths in Table 6.1 have been obtained assuming that helicity 2 is the only contribution. The average of the published results is (Table 6.2):

$$\Gamma(f \to \gamma\gamma) = 2.95 \pm 0.3 \text{ keV} \qquad (\lambda = 2 \text{ assumed}).$$

6.4.3   The Two-Photon Coupling of the $A_2(1320)$

The $A_2(1320)$ is the isovector partner of the f in the 2^{++} multiplet. The two-photon production of the A_2 was first observed by the Crystal Ball collaboration /108/ in the decay mode

$$A_2 \to \pi^0 \eta, \tag{6.37}$$

which has a branching ratio of 14.5±1.2%. Both the π^0 and the η were detected via their decays into two photons. In the same plot (Fig.6.18) which shows the $\pi^0\pi^0$ enhancement due to the f resonance, one also sees an enhancement in the $\pi^0\eta$ region. The background subtracted $\pi^0\eta$ mass distribution is plotted in Fig.6.21. The A_2 signal is clearly visible. The statistics are too poor to show whether there is any additional resonance production (e.g. the $\delta(980)$). The $\gamma\gamma$ width has been determined with the assumption that the A_2 is produced in a helicity 2 state (decay angular distribution $\sim \sin^4\vartheta^*$). The result is included in Table 6.1.

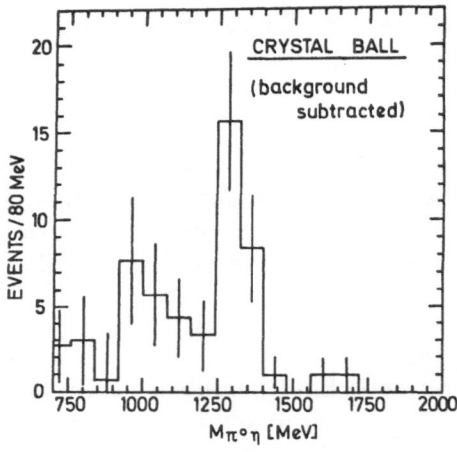

Fig.6.21.   The $\pi^0\eta$ mass spectrum measured by the Crystal Ball.

For magnetic detectors, the decay mode $A_2 \to \rho\pi$ (branching ratio: 70.1±2.2%) is preferable. CELLO /78/ and JADE /80/ observed the A_2 in the $\pi^0\pi^+\pi^-$ final state:

$$\gamma\gamma \to A_2 \to \rho^\pm\pi^\mp \to \pi^0\pi^+\pi^-. \tag{6.38}$$

The $\rho^0\pi^0$ decay mode is forbidden by isospin conservation.

The combined angular distributions for the A_2 decay and the subsequent decay of the ρ are:

$$|\psi(\lambda=0)|^2 \sim \sin^2\vartheta_\rho \cos^2\vartheta_\rho \sin^2\vartheta_\pi \sin^2\varphi_\pi$$

$$|\psi(\lambda=1)|^2 \sim \sin^2\vartheta_\pi \left[(2\cos^2\vartheta_\rho - 1)^2 \sin^2\varphi_\pi + \cos^2\vartheta_\rho \cos^2\varphi_\pi\right] \tag{6.39}$$

$$|\psi(\lambda=2)|^2 \sim \sin^2\vartheta_\rho \sin^2\vartheta_\pi (\cos^2\vartheta_\rho \sin^2\varphi_\pi + \cos^2\varphi_\pi).$$

The decay of the A_2 is described by the polar angle ϑ_ρ of the ρ in the $\gamma\gamma$ center of mass system, with the z axis in the $\gamma\gamma$ direction. An overall azimuthal dependence cannot be measured without the detection of the scattered electrons. The angles ϑ_π, φ_π describe the decay of the ρ in the ρ helicity system, which is the ρ rest system with the z axis pointing into the direction of flight of the ρ; φ_π is measured with respect to the plane defined by the ρ and $\gamma\gamma$ directions. The assumption of helicity 2 dominance can be tested by analysing the angular correlations in the three pion final state. First preliminary results by JADE indicate that $\lambda=2$ indeed dominates /80/.

The CELLO collaboration observed the A_2 in the decay channel (6.38) by detecting only one of the photons from the π^0 decay, i.e. they observed the A_2 in the same $\pi^+\pi^-\gamma$ mass plot as the η' (Fig.6.22) /78/. Loosing one low energy photon does not significantly distort the A_2 mass distribution; requiring the detection of both photons would decrease the efficiency by a factor ~5. The $\gamma\gamma$ width has been derived with the assumption $\lambda=2$ and is given in Table 6.1. The average of the published results from Crystal Ball and CELLO is (Table 6.2):

$$\Gamma(A_2 \to \gamma\gamma) = 0.79 \pm 0.23 \text{ keV} \qquad (\lambda=2 \text{ assumed}).$$

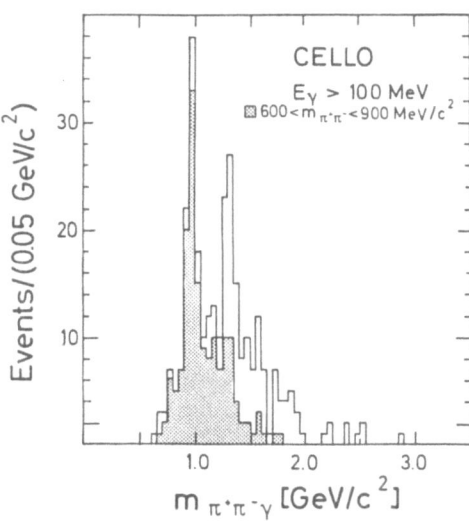

Fig.6.22. The $\pi^+\pi^-\gamma$ mass spectrum measured by CELLO: Besides the η' peak, a signal is also seen at the mass of the A_2, which is due to the $\pi^+\pi^-\pi^0$ decay mode (one photon from the π^0 decay is missing). Requiring the $\pi^+\pi^-$ mass to lie in the ρ band (dotted histogram) does not affect the η' signal but reduces the A_2 contribution.

6.4.4 The Two-Photon Coupling of the f'(1515)

The third flavour neutral member of the 2^{++} multiplet is the f'(1515). The dominance of the decay into $K\bar{K}$ indicates that the quark content is mainly $s\bar{s}$ /67/. Thus it appears that the f - f' system is a nearly ideal mixture of SU(3) states, with the f being mainly $u\bar{u}+d\bar{d}$ and the f' mainly $s\bar{s}$. According to the quark model, the $\gamma\gamma$ width of the nonet member with dominant $s\bar{s}$ content should be highly sensitive to deviations from ideal mixing because of the difference in the charges of the s and u quarks.

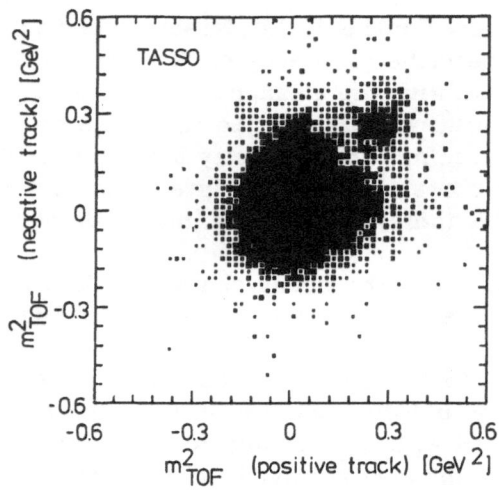

Fig.6.23. Identification of charged kaon pairs by time-of-flight (TASSO): Square of the mass as calculated from time-of-flight for the negative track versus the same quantity for the positive track in events with 2 charged tracks detected.

Upper limits for $\Gamma(f'\to\gamma\gamma)\cdot B(f'\to K\bar{K})$ were given in /24/. The TASSO collaboration reported direct evidence for the reaction $\gamma\gamma\to f'(1515)$ /110/, where the f' was observed in the two decay channels

$$f'\to K^+K^- \qquad \text{and} \qquad f'\to K^0_s K^0_s. \tag{6.40}$$

Using isospin symmetry the relative contributions from different K-pairings to all $K\bar{K}$ decays of a C-even resonance are found to be:

$$K^+K^- : K^0_s K^0_s : K^0_L K^0_L = 0.5 : 0.25 : 0.25. \tag{6.41}$$

The TASSO group selected events with two charged K's from a sample of events with two tracks of opposite charge by means of time-of-flight measurements. In Fig.6.23 the square of the mass calculated from the time-of-flight for the negative charged track is plotted versus the same quantity for the positive charged track. A distinct cluster is observed at the masses of K^+ and K^-. The mass spectrum of identified K^+K^- pairs (Fig.6.24) shows a peak in the region of the f'. The enhancement below the f' is probably due to f and A_2 decays into K^+K^- and due to continuum production of K^+K^- (similar to the $\pi^+\pi^-$ continuum discussed above).

A peak in the f' region has also been observed in the invariant mass of $K^0_s K^0_s$ pairs (Fig.6.25). The K^0_s's were detected via their decays into $\pi^+\pi^-$. Fig.6.25 also shows the background determined from sidebands in the two-dimensional plot of $\pi^+\pi^-$ versus $\pi^+\pi^-$ mass combinations.

The contribution of a resonance to the measured $K\bar{K}$ mass spectra depends on the product of the $\gamma\gamma$ width and the decay branching ratio into $K\bar{K}$. For the f and A_2 resonances this product is probably of similar magnitude as for the f' resonance. The branching ratios into $K\bar{K}$ are: $2.9\pm0.2\%$ for the f, $4.8\pm0.5\%$ for the A_2 and only known to be "dominant" for the f' /72/. Because of the small mass differences between these three states possible interferences have to be taken into account. The contributions of the 2^{++} nonet to the cross section for $\gamma\gamma\to K\bar{K}$ can be expressed as a coherent sum of the f, A_2 and f' amplitudes, which reads in the case of ideal mixing /111/:

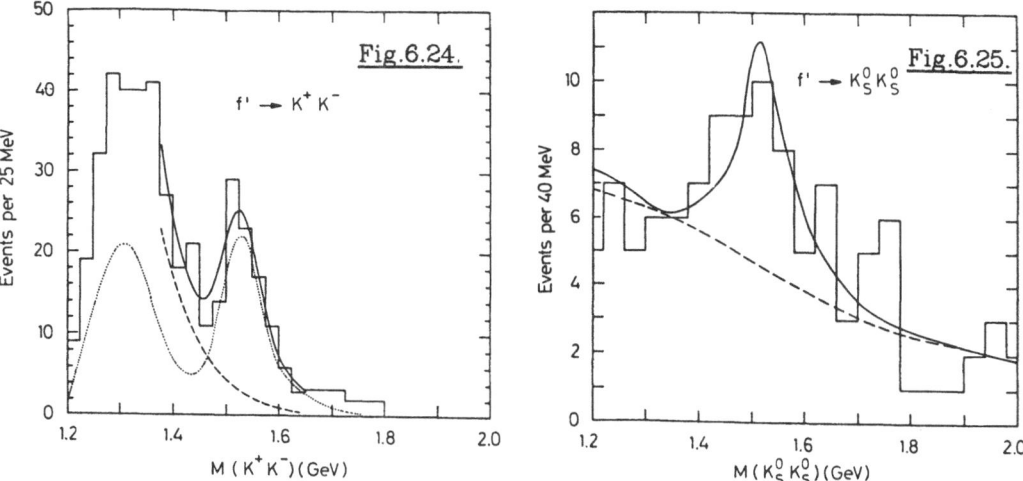

<u>Fig.6.24.</u> Mass spectrum of charged kaon pairs (TASSO): The dotted curve is the contribution from the interfering f, A_2 and f' resonances. The background from continuum production of K^+K^- is estimated by the dashed curve. The solid curve is the result of a fit with resonances and background.

<u>Fig.6.25.</u>   Mass spectrum of neutral kaon pairs (TASSO): The solid curve is the result of a fit with interfering f, A_2 and f' resonances and background. A background from continuum production of $K^0\bar{K}^0$ has been neglected. The backgound from non-$K^0\bar{K}^0$ has been determined from sidebands (dashed curve).

$$\sigma_{\gamma\gamma\to K\bar{K}}(W_{\gamma\gamma}) = (40\pi/W_{\gamma\gamma}^2) \; \Big| [\Gamma(f\to\gamma\gamma)\cdot B(f\to K\bar{K})]^{1/2}\cdot BW(f)$$
$$\pm \; [\Gamma(A_2\to\gamma\gamma)\cdot B(A_2\to K\bar{K})]^{1/2}\cdot BW(A_2)$$
$$+ \; [\Gamma(f'\to\gamma\gamma)\cdot B(f'\to K\bar{K})]^{1/2}\cdot BW(f')\Big|^2 \; . \qquad (6.42)$$

$BW(R)=M\sqrt{\Gamma}/[M^2-W_{\gamma\gamma}^2-iM\Gamma]$ is a relativistic Breit Wigner amplitude, where M is the mass and Γ is the energy dependent width of the resonance R. $B(R\to K\bar{K})$ is the $K\bar{K}$ branching ratio of the resonance R, and $\Gamma(R\to\gamma\gamma)$ its $\gamma\gamma$ partial width. According to SU(3), the '+' sign of the second term applies to the K^+K^- final state; the '–' sign applies to $K^0\bar{K}^0$, where the interference between the isovector (A_2) and the isoscalars (f, f') is destructive /111/. The change of sign is due to the fact that charged K's contain u quarks whereas neutral K's contain d quarks; the phase of the u quarks is opposite (equal) to that of the d quarks in the isovector (isoscalar) wave function. The cross section obtained from (6.42) is shown in Fig.6.26 for the final states K^+K^- (solid curve) and $K^0\bar{K}^0$ (dashed curve) using measured parameters for the three resonances.

The most distinctive feature in this plot is the large f and A_2 contribution to the K^+K^- mode, whereas their contribution to the $K^0\bar{K}^0$ mode is small. These different interference patterns are consistent with the data.

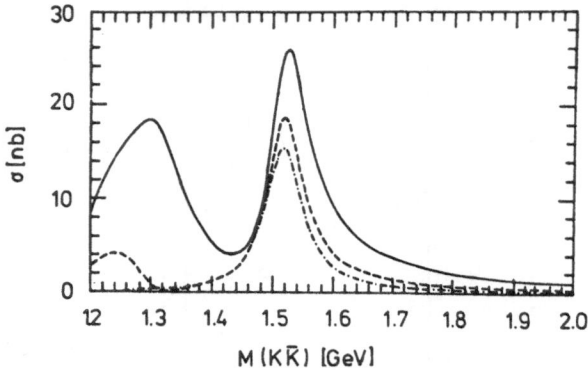

Fig.6.26.   Interference   pattern   for $\gamma\gamma{\rightarrow}f$, A_2,   f' in the K^+K^- (solid
curve) and $K^0\bar{K}^0$ (dashed curve) final states.   The dashed-dotted curve
shows the f' resonance without interference.

However, the uncertainty in the K^+K^- continuum contribution and the limited statistics in the $K^0_sK^0_s$ channel do not allow an independent test of the
validity of this ansatz.

The final result for $\Gamma(f'{\rightarrow}\gamma\gamma){\cdot}B(f'{\rightarrow}K\bar{K})$ has been obtained by fitting a sum
of resonances and a background contribution to each of the two mass
spectra independently. The resonances have been described by the interference pattern given by (6.42) assuming helicity 2 production. For the
K^+K^- channel the non-resonant background has been taken into account
by a phenomenological parametrization (dashed curve in Fig.6.24). For the
$K^0_sK^0_s$ channel the shape of the background was determined from the
sidebands and its size was adjusted by the fit (dashed curve in Fig.6.25);
non-resonant $K^0_sK^0_s$ production was neglected. The results of the fits to
the K^+K^- and $K^0_sK^0_s$ mass distributions are in agreement with one another
and lead to the combined result:

$$\Gamma(f'{\rightarrow}\gamma\gamma){\cdot}B(f'{\rightarrow}K\bar{K}) = 0.11\pm0.02\pm0.04 \text{ keV}.$$

Since the branching ratio for f' decays into $K\bar{K}$ is not yet known, the $\gamma\gamma$
width of the f' could not be extracted.

6.4.5   Other 2^{++} Mesons?

Up to now there is no evidence for a coupling of other charge and
flavour neutral 2^{++} mesons to two photons in the mass range of 1 to 2 GeV.
Special attention has been paid to the glueball candidate state $\Theta(1640)$ dis-
covered in radiative J/ψ decays /112/. This state is likely to be a $J^{PC}=2^{++}$
state. The observed decay modes are:

$$\Theta{\rightarrow}\eta\eta \text{ /112/}, \qquad \Theta{\rightarrow}K\bar{K} \text{ /91/}, \qquad \Theta{\rightarrow}\rho^0\rho^0 \text{ /113/},$$

though it has not yet been established that all decays come from the same
object. The TASSO collaboration gave upper limits on the products of the

$\gamma\gamma$ width times the branching ratios into $\rho^\circ\rho^\circ$ /114/ (see also Sect.6.7.1) and $K\bar{K}$ /110/:

$\Gamma(\Theta\to\gamma\gamma)\cdot B(\Theta\to\rho^\circ\rho^\circ) < 1.2$ keV ,   (95% c.l.)

$\Gamma(\Theta\to\gamma\gamma)\cdot B(\Theta\to K\bar{K}) < 0.3$ keV ,   (95% c.l.).

The Crystal Ball group derived a limit for two-photon produced Θ's decaying into $\eta\eta$ /115/:

$\Gamma(\Theta\to\gamma\gamma)\cdot B(\Theta\to\eta\eta) < 0.3$ keV ,   (95% c.l.).

6.4.6 Discussion of the Results for Tensor Mesons

In the following we want to discuss the implications of the measured $\gamma\gamma$ widths of the tensor mesons for their quark model assignments. As mentioned before no absolute prediction for these $\gamma\gamma$ widths is rigorously possible, which is well demonstrated by the wide range of predicted values available. We therefore rather concentrate on the examination of the ratios of the $\gamma\gamma$ widths, which are on a much better theoretical footing. From the results in Table 6.2 we get:

$$\frac{\Gamma(A_2\to\gamma\gamma)}{m_{A2}{}^3} : \frac{\Gamma(f\to\gamma\gamma)}{m_f{}^3} : \frac{\Gamma(f'\to\gamma\gamma)}{m_{f'}{}^3} = 9 : (37\pm11) : (0.8\pm0.4)/B(f'\to K\bar{K}).$$

Assuming that $m_T{}^3$ is the proper phase space correction, i.e. that (6.31) holds (see Sect.6.4.1), ideal mixing in the 2^{++} nonet would give for these ratios 9 : 25 : 2 according to (6.29). From the ratio of the $\gamma\gamma$ widths of A_2, f and f' one can determine the mixing angle in the same way as was done for the pseudoscalars. Using (6.12) the TASSO group finds for $B(f'\to K\bar{K}) > 0.5$ the allowed range for the mixing angle Θ /110/:

25.4° < Θ < 34.7° (95% c.l.).

This result indicates some deviation from the ideal mixing angle of 35.3° and is in agreement with the value found from the Gell-Mann/Okubo mass formula Θ=28±3° /72/.

The constraint imposed on the non-strange quark content in the f' is demonstrated in Fig.6.27a. In this plot the theoretical ratio $\Gamma(f'\to\gamma\gamma)/\Gamma(f\to\gamma\gamma)$ as a function of the mixing angle Θ is compared to the experimental bounds which hold for all values of $B(f'\to K\bar{K})$ between 50% and 100%. The u,d quark content is given by the deviation from ideal mixing and is constrained by the data to be less than 3% (95% c.l.) /110/. It is interesting to note that for $\Theta\approx20°$, corresponding to a light quark content of about 8%, the $\gamma\gamma$ width of the f' would vanish. Figure 6.27b shows the same plot for $\Gamma(A_2\to\gamma\gamma)/\Gamma(f\to\gamma\gamma)$. Within the errors, the experimental result for this ratio is consistent with the SU(3) prediction.

Models which include a 2^{++} gluonic bound state (usually taken as the $\Theta(1640)$) and which account for the possibility of a more general mixing

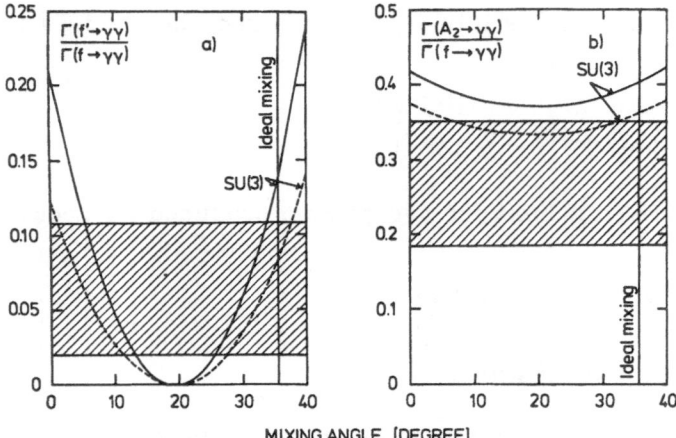

<u>Fig.6.27.</u>   SU(3) prediction for the ratios of $\gamma\gamma$ widths of tensor mesons as a function of the mixing angle Θ. : a) $\Gamma(f'\rightarrow\gamma\gamma) : \Gamma(f\rightarrow\gamma\gamma)$; b) $\Gamma(A_2\rightarrow\gamma\gamma) : \Gamma(f\rightarrow\gamma\gamma)$. The full curves are the predictions assuming a phase space correction $\sim 1/m^3$, while the dashed curves do not contain this correction. The shaded areas are the one standard deviation intervals allowed by the experimental results. In a) this interval applies for $50\% < B(f'\rightarrow K\overline{K}) < 100\%$.

between quark states and gluonic states have been discussed by many authors /117/. The couplings of a pure glueball to normal hadrons should be flavour independent (see also discussion in /91/). Experimentally, this does not hold for the Θ, because the decay branching ratio into $\pi\pi$ is found to be much smaller than those into $\eta\eta$ and $K\overline{K}$ /91/. Many authors nevertheless tried to defend the glueball hypothesis and constructed models explaining the observed branching ratios. For example, in the model described in /118/ the decay branching ratios of the Θ are explained by invoking mixing with $q\overline{q}$ states (f, f') which provides the desired suppression of the $\pi\pi$ decay, if the parameters are properly chosen. Using the constraints from the masses and measured decay branching ratios of the f and f', this scheme leads to the prediction $\Gamma(f'\rightarrow\gamma\gamma) < 0.01$ keV, which is clearly in contradiction to the TASSO measurements. Since the prediction of a nearly vanishing $\gamma\gamma$ width of the f' is forced by the assumption that the $\Theta(1640)$ is a 2^{++} gluonic state, one has to conclude that in this special model the glueball interpretation of the $\Theta(1640)$ is ruled out by the TASSO results.

6.5 The Two-Photon Coupling of Scalar Mesons

The experimental situation of the scalar meson nonet (0^{++}) is very obscure. No scalar states have been found which fit readily into the simple quark model, in contrast to the apparently nice behaviour of the tensor meson nonet. (Note that both spin-parity states belong to the same P-wave triplet in the quark model). Various explanations for the failure of the simple quark model predictions have been given /67/:

- an anomalous $K\bar{K}$ threshold behaviour,
- mixing with $qq\bar{q}\bar{q}$,
- effects from QCD vacuum fluctuations with the same quantum numbers,
- uncertainties in the phase shift analyses.

Candidates for the flavour neutral members of the scalar nonet are the isoscalars ε(~800 and/or ~1300 ?) and S*(975) and the isovector δ(980). Our knowledge on the isoscalars comes mainly from $\pi\pi$ and KK phase shift analyses. Because the inelasticity for $\pi\pi$ scattering is small below ~1 GeV one can assume that in this region the ($\gamma\gamma\to\pi\pi$) phase shift is the same as in the elastic $\pi\pi$ channel (Watson theorem) /12/. Note that resonance shapes as seen in $\gamma\gamma\to\pi\pi$ and $\pi\pi\to\pi\pi$ may be different. The $\gamma\gamma\to\pi\pi$ cross section can be large even if the phase shift is zero as is the case for the Born term near threshold (Chap.5).

An experiment at the DCI storage ring measured the two photon production of $\pi^+\pi^-$ from threshold up to about 0.7 GeV /41/ (see Sect.5.1). They observe within the detector acceptance a cross section of 69±15 pb where 34 pb are expected from a simple Born term. This 2.3 standard deviation effect has been taken as a hint for the presence of a broad scalar resonance ε(800). A model including an ε(800) and an S*(975) was able to describe the data /50/. This is an extremely interesting result. However, since the statistics of this experiment are not overwhelming and the systematic uncertainties in deriving the number of pion pairs (subtraction of the tails from e^+e^- and $\mu^+\mu^-$ production, see Fig.4.6) are not negligible, a second experiment is highly desirable. (Note that the Mark II group does not observe a significant deviation from the Born term down to about 0.5 GeV as can be seen in Fig.6.15).

The TASSO group investigated the $\pi^+\pi^-$ channel in connection with the f analysis. For the region above the f peak, where the experiment is sensitive to S wave $\pi^+\pi^-$ production, they derived an upper limit for the $\gamma\gamma$ coupling to a scalar resonance with a mass between 1.3 and 1.5 GeV /44/:

$$\Gamma(\varepsilon\to\gamma\gamma)\cdot B(\varepsilon\to\pi^+\pi^-) < 1.5 \text{ keV (95\% c.l.)}.$$

This limit is relevant in connection with a recent result of a $\pi\pi$ and KK phase shift analysis which yields for the ε a mass of 1.425±0.015 GeV and a width of 0.165±0.030 GeV /119/.

Another possible member of the 0^{++} nonet is the S*(975) which was found as a pole in the isoscalar S wave of $\pi\pi$ and KK scattering amplitudes. The Crystal Ball collaboration gave an upper limit from the analysis of the $\pi^\circ\pi^\circ$ mass spectrum /108/:

$$\Gamma(S^*\to\gamma\gamma)\cdot B(S^*\to\pi\pi) < 0.8 \text{ keV (95\% c.l.)}.$$

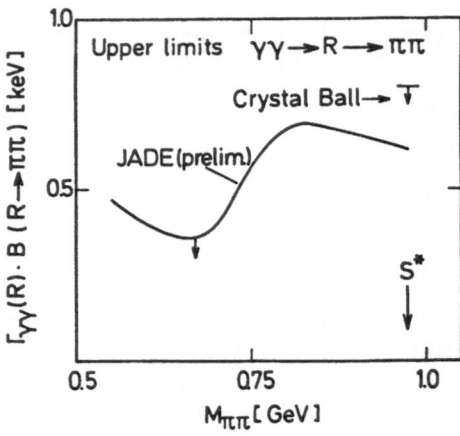

Fig.6.28. Upper limits (95% c.l.) for $\gamma\gamma \to R \to \pi\pi$: R is a narrow resonance ($\Gamma < 50$ MeV) with a decay branching $B(R \to \pi\pi)$ into pions. The limits from JADE have been derived from the $\pi^0\pi^0$ mass distribution in Fig.6.20.

The JADE collaboration derived (preliminary) upper limits for narrow resonances below 1 GeV decaying into $\pi^0\pi^0$, see Fig.6.28 /80/. Their upper limit for the $\gamma\gamma$ width of the S* is obtained by using the $\pi\pi$ decay branching ratio of 78 ± 3 % from /72/:

$$\Gamma(S^* \to \gamma\gamma) < 0.8 \text{ keV (95\% c.l.)}.$$

No experimental results have been reported yet on the $\gamma\gamma$ coupling of the possible isovector member of the scalar nonet, the $\delta(980)$.

In most of the theoretical estimates the $\gamma\gamma$ width of the ε is predicted to be comparable to the f width or even larger /96 to 104/. An exception is the model of /120/ which predicts a suppression of the $\gamma\gamma$ widths of all 0^{++} $q\bar{q}$ states. According to this model, the observation of a large $\gamma\gamma$ coupling for a 0^{++} state would indicate that this state is not purely $q\bar{q}$. In /121/ it was pointed out that the $\gamma\gamma$ width of the low-lying isoscalar, ε or σ, can be derived from a low energy theorem similarly as in the case of the pseudoscalars yielding the expression:

$$\Gamma(\sigma \to \gamma\gamma) = \frac{m_\sigma^3}{16\pi}\left(\frac{\alpha R}{8\pi^2 f_\sigma}\right)^2 \tag{6.43}$$

where $R = \sigma(e^+e^- \to \text{hadrons})/\sigma(e^+e^- \to \mu^+\mu^-)$ and $f_\sigma \approx 100\text{-}150$ MeV. For $m_\sigma \approx 700$ MeV it was estimated /122/:

$$\Gamma(\sigma \to \gamma\gamma) \approx 6 \text{ keV}.$$

Another estimate for the $\gamma\gamma$ coupling of the scalar mesons can be obtained from a sum rule, which relates the cross sections for $\gamma\gamma$ helicities $\lambda=0$ and $\lambda=2$ /123/:

$$\int_0^\infty \frac{dW_{\gamma\gamma}^2}{W_{\gamma\gamma}^2}\left(\sigma^{\lambda=0}(W_{\gamma\gamma}) - \sigma^{\lambda=2}(W_{\gamma\gamma})\right) = 0. \tag{6.44}$$

It was shown that pointlike QED processes, like the quark box diagram and the $\pi^+\pi^-$ Born term, satisfy this relation /103, 124/. Saturating the sum rule with the low lying scalar, pseudoscalar and tensor states gives:

$$\sum_S \frac{\Gamma_{S\gamma\gamma}}{m_S{}^3} + 5 \sum_T \frac{\Gamma_{T\gamma\gamma}(\lambda=0)}{m_T{}^3} \approx 5 \sum_T \frac{\Gamma_{T\gamma\gamma}(\lambda=2)}{m_T{}^3} - \sum_P \frac{\Gamma_{P\gamma\gamma}}{m_P{}^3}. \qquad (6.45)$$

With the measured $\gamma\gamma$ widths of the pseudoscalars and tensors (Table 6.2) the right hand side (RHS) of (6.45) gives:

RHS = -1.9 ± 0.9 keV/GeV³.

At the level of two standard deviations, the value of RHS is inconsistent with the requirement that the left hand side be positive definite. We checked that the narrow resonance approximation does not cause this problem. However, the interference of the f resonance with the $\pi^+\pi^-$ continuum changes the f contribution to RHS by about 50% if the $\pi^+\pi^-$ continuum is estimated by the Born term. With this correction we get:

RHS = 1.4 ± 1.1 keV/GeV³   (including interference effects for
                                        the f in the $\pi^+\pi^-$ channel).

If the left hand side of (6.45) is dominated by one scalar resonance ε, we get for the $\gamma\gamma$ width:

$\Gamma(\varepsilon \rightarrow \gamma\gamma)$ = 0.5 ± 0.4 keV   for M_ε = 0.7 GeV
                     = 3.1 ± 2.4 keV   for M_ε = 1.3 GeV.

6.6   Remarks on the Two-Photon Coupling of Charmonium States

Potential models /68/ and QCD sum rules /125/ have been very successfully used to describe heavy quarkonia. This includes the calculations of energy levels, leptonic and radiative widths and to some extent also hadronic widths. The hadronic decays of heavy quarkonia are assumed to proceed in lowest order via the smallest number of gluons possible: two gluons for C-even states with spin J≠1 and three otherwise (decays of J=1 states via two real gluons are not allowed, but the gluons may not always be real in these decays). In lowest order the two gluon decay (with subsequent fragmentation into hadrons) is described by the same amplitude as the $\gamma\gamma$ decay if one uses the strong coupling constant α_s instead of the fine structure constant α. Thus the ratio of the hadronic width to the $\gamma\gamma$ width for a C=+1 state is:

$$\Gamma_{\gamma\gamma}/\Gamma_{had} = \Gamma_{\gamma\gamma}/\Gamma_{gg} = 9/2 \cdot e_q{}^4 (\alpha/\alpha_s)^2. \qquad (6.46)$$

Therefore, by measuring $\Gamma_{\gamma\gamma}$ and Γ_{had} one can determine α_s or, turning the argument around, this relation allows in principle a test of the various assumptions made, e.g. the number of gluons involved, the higher order corrections etc..

To get a prediction for the $\gamma\gamma$ width of charmonium states one cannot rely on SU(4) symmetry since the symmetry is strongly broken by the mass of the charm quark. One can, however, relate the $\gamma\gamma$ width of a singlet state (like the η_c) to the leptonic width of the corresponding triplet state (like the J/ψ). Both depend on the wave function $f(0)$ at the origin, which, to first approximation, should be equal for the singlet and triplet states:

$$\Gamma(^1S_0\to\gamma\gamma)/\Gamma(^3S_0\to l^+l^-) = \frac{3\cdot e_q^4|f(0)|^2/m_q^2}{e_q^2|f(0)|^2/m_q^2} \left\{\begin{array}{l} =4/3 \text{ for } q=c \\ =1/3 \text{ for } q=b. \end{array}\right. \tag{6.47}$$

Using the measured leptonic widths of the J/ψ, ψ' and Υ yields the following estimates:

$\Gamma(\eta_c\to\gamma\gamma) \approx 6.2 \text{ keV}$

$\Gamma(\eta_c'\to\gamma\gamma) \approx 2.6 \text{ keV}$

$\Gamma(\eta_b\to\gamma\gamma) \approx 0.4 \text{ keV}.$

Similar results were obtained using QCD sum rules /126/. However, in a recent paper /127/ large differences between the singlet and triplet wave functions were found, which would require large corrections to the numerical values obtained from (6.47). This finding is supported experimentally by the measurement of the branching ratio $B(J/\psi\to\gamma\eta_c)$ /135/, which turns out to be much smaller than expected if the overlap of the singlet and triplet wave functions were complete. In the following formula, obtained from dispersion relations in /128/ (see also the discussion in /91/), the assumption of equal wave functions was not made:

$$\Gamma(J/\psi\to\gamma\eta_c) = \alpha(16/9)E_\gamma^3[\Gamma(\eta_c\to\gamma\gamma)/\Gamma(J/\psi\to e^+e^-)](M_{J/\psi}/M_{\eta_c}^3)(1-0.28\cdot\alpha_s). \tag{6.48}$$

E_γ is the photon energy of the J/ψ - η_c transition. With the measured values for $\Gamma(J/\psi\to\gamma\eta_c)$ and $\Gamma(J/\psi\to e^+e^-)$ and with $\alpha_s = 0.3$ the result for the $\gamma\gamma$ width of the η_c is much lower than the one obtained from (6.47):

$\Gamma(\eta_c\to\gamma\gamma) = 1.6\pm0.8 \text{ keV}.$

No $\gamma\gamma$ decay or $\gamma\gamma$ production of the $\eta_c(2980)$ has been observed up to now. The Crystal Ball group obtained the following upper limit from the study of 3γ decays of the J/ψ /129/:

$B(J/\psi\to\gamma\eta_c)\cdot B(\eta_c\to\gamma\gamma) < 1.6\cdot10^{-5}$, 90% c.l..

Using the branching ratio $B(J/\psi\to\gamma\eta_c)$ and the total width of the η_c one gets /129/:

$\Gamma(\eta_c\to\gamma\gamma) < 20 \text{ keV}$, 90% c.l..

The expected value for the $\gamma\gamma$ width of the η_c is not too small, but it will be difficult to observe the η_c in two-photon collisions for the following reasons: 1) The production rate of η_c's in colliding beams is small compared to lower lying resonances because for a given $\gamma\gamma$ width the rate drops like M_R^{-3} (M_R = mass of the resonance) and 2) the detection efficiency may be small because the η_c has probably no dominant decay mode and a relatively high average multiplicity of the decay products. The decay

Table 6.3. Measured decay modes of the η_c.

Decay Mode	$B(\eta_c \to X)$ [%]	Experiment
$\eta_c \to K^0_s K^{\pm} \pi^{\mp}$	5.4 +3.3 -2.4	Mark II /131/
$\eta_c \to K^+ K^- \pi^+ \pi^-$	1.4 +2.1 -0.9	Mark II /131/
$\eta_c \to \pi^+ \pi^- \pi^+ \pi^-$	2.0 +1.5 -0.9	Mark II /131/
$\eta_c \to p\bar{p}$	0.29 +0.30 -0.16	Mark II /131/
$\eta_c \to \pi^+ \pi^- p\bar{p}$	<2.3 (90% c.l.)	Mark II /131/
$\eta_c \to \eta \pi^+ \pi^-$	2.6 +1.8 -1.7	Crystal Ball /132/
$\eta_c \to K^+ K^- \pi^0$	<1.7 (90% c.l.)	Crystal Ball /129/
$\eta_c \to \gamma\gamma$	<0.18 (90% c.l.)	Crystal Ball /129/
$\eta_c \to \varphi\varphi$	1.0 +0.4 -0.4	Mark III /133/

modes of the η_c which have been observed in radiative decays of the J/ψ by the Mark II /131/, Crystal Ball /132/ and Mark III /133/ experiments are listed in Table 6.3. These channels add up to a combined branching ratio of not more than about 15 %. All seen modes, with the exception of the $\eta\pi^+\pi^-$ decay, involve only charged particles in the final state, which is convenient for magnetic detectors.

The TASSO and JADE groups searched for the η_c in various decay channels and obtained the upper limits listed in Table 6.4. Considering the measured branching ratios, none of the limits is really stringent. The numbers rather reflect the sensitivity the experiments have reached with an collected integrated luminosity of about 80 pb^{-1}. Whether the two-photon production of the η_c will be seen in the near future at PETRA or PEP will depend crucially on whether the actual $\gamma\gamma$ width is closer to the higher or to the lower end of the theoretical estimates discussed above.

Estimates for the $\gamma\gamma$ widths of the P-wave states have to rely on the gluon counting rule, which may not give more than the order of magnitude (some keV for the $\chi_0(3415)$ and $\chi_2(3555)$ states). More reliable predictions are obtained from the quark spin structure for the ratios of the $\gamma\gamma$ widths within the P-wave triplet /88/:

$$\Gamma(\chi_0 \to \gamma\gamma)/\Gamma(\chi_2 \to \gamma\gamma) = 15/4. \tag{6.49}$$

Table 6.4. Upper limits for two photon production of the η_c.

Decay X	$\Gamma(\eta_c \to \gamma\gamma) \cdot B(\eta_c \to X)$ [keV]		Ref.	$B(\eta_c \to X)$ [%]
all	< 20	90% c.l.	/129/	
$p\bar{p}$	< 0.32	95% c.l.	/55/	0.29 +0.30 -0.16
$K\bar{K}\pi$	< 27	95% c.l.	/80/	16.2 +9.9 -7.2
$\pi^+\pi^-\pi^+\pi^-$	< 0.7	95% c.l.	/80/	2.0 +1.5 -0.9
$\pi^+\pi^-\pi^0\pi^0$	< 4.2	95% c.l.	/80/	(1-2)·(2.0 +1.5 -0.9)
$\eta\pi^+\pi^-$	< 2.3	95% c.l.	/80/	2.6 +1.8 -1.7

The Crystal Ball collaboration looked for the two-photon decays of these two states in radiative transitions from the ψ':

$$\psi' \to \gamma\chi \to \gamma\gamma\gamma. \tag{6.50}$$

Preliminary results are /134/:

$$B(\chi_0 \to \gamma\gamma) < 5\cdot 10^{-4},\ 90\%\ \text{c.l.}$$
$$B(\chi_2 \to \gamma\gamma) = (6\pm 2)\cdot 10^{-4}. \tag{6.51}$$

Combining these results with the hadronic branching ratios from /72/ [with $B(\chi \to \text{hadrons}) = 1 - B(\chi \to \gamma J/\psi)$] one obtains the following ratios, which can be compared with the theoretical prediction in (6.46):

$$\Gamma(\chi_0 \to \gamma\gamma)/\Gamma(\chi_0 \to gg) < 5\cdot 10^{-4}$$
$$\Gamma(\chi_2 \to \gamma\gamma)/\Gamma(\chi_2 \to gg) = (7.1\pm 2.4)\cdot 10^{-4}. \tag{6.52}$$

From the gluon counting rules we expect that this ratio should be the same for both states. This is not ruled out by the data. Assuming α_s to be between 0.2 and 0.4, the ratios in (6.52) are in agreement with relation (6.46). The branching ratios in (6.51) can be converted to $\gamma\gamma$ widths by using the measured total widths from /135/:

$$\Gamma(\chi_0 \to \gamma\gamma) < 10\ \text{keV},\ 90\%\ \text{c.l.}$$
$$\Gamma(\chi_2 \to \gamma\gamma) = 1.8\pm 1.3\ \text{keV}.$$

The ratio of these values is not inconsistent with (6.49). It may be interesting to note that the corresponding ratio for the total hadronic widths of the χ_0 and the χ_2 is experimentally much larger than expected /135/.

6.7 The Four Pion Final State and Vector Meson Pair Production

The cross section for $\rho^\circ\rho^\circ$ production by two photons, first observed by the TASSO collaboration /136/, shows a large enhancement near the $\rho^\circ\rho^\circ$ threshold. So far, this is the only effect observed in two-photon experiments which had not been predicted, not even qualitatively. Thus, this observation initiated a lot of experimental activities and theoretical speculations. In the following we report on the detailed investigations of the $\rho^\circ\rho^\circ$ production in the threshold region. Clues for the interpretation of the $\rho^\circ\rho^\circ$ enhancement may come from studying the production of other vector meson pairs. First results have been reported on the production of $\rho^+\rho^-$, $\rho^\circ\omega$, $\omega\omega$.

6.7.1 $\rho^\circ\rho^\circ$ Production by Two Photons Near Threshold

The TASSO collaboration observed in the channel

$$\gamma\gamma \to \rho^\circ\rho^\circ \tag{6.53}$$

a large enhancement near threshold /136/. These measurements have been confirmed by other experiments /137, 138/. The TASSO group recently published a new analysis based on about 20 times more events than in the first analysis /114/.

The cross section for $\gamma\gamma \to \rho^\circ\rho^\circ$ near the $\rho^\circ\rho^\circ$ threshold (Fig.6.29) is about one order of magnitude larger than expected from a simple VMD-Regge estimate taking only the Pomeron exchange into account. In the VMD model two-photon scattering is related to $\rho^\circ\rho^\circ$ scattering:

$$d\sigma(\gamma\gamma \to \rho^\circ\rho^\circ)/dt = (\alpha\pi/\gamma_\rho^2)^2 \cdot (p^*/k^*)^2 \cdot d\sigma(\rho^\circ\rho^\circ \to \rho^\circ\rho^\circ)/dt \qquad (6.54)$$

with $\alpha\pi/\gamma_\rho^2 = 2.8 \cdot 10^{-3}$ and k^*, p^* the center of mass momenta of the γ's and ρ's respectively. Applying the optical theorem and relating the total $\rho^\circ\rho^\circ$ cross section via the additive quark model to the total pp cross section at high energies, one finds (see curve in Fig.6.29):

$$\sigma^{\text{pomeron}}(\gamma\gamma \to \rho^\circ\rho^\circ) \approx (p^*/k^*)^2 \cdot 35 \text{ nb.} \qquad (6.55)$$

Such an estimate may be too simple minded, because only pomeron exchanges are taken into account by relating high energy cross sections. A different approach in the spirit of the VMD model has been taken in /139/. The authors use factorization to relate the $\gamma\gamma$ cross section to

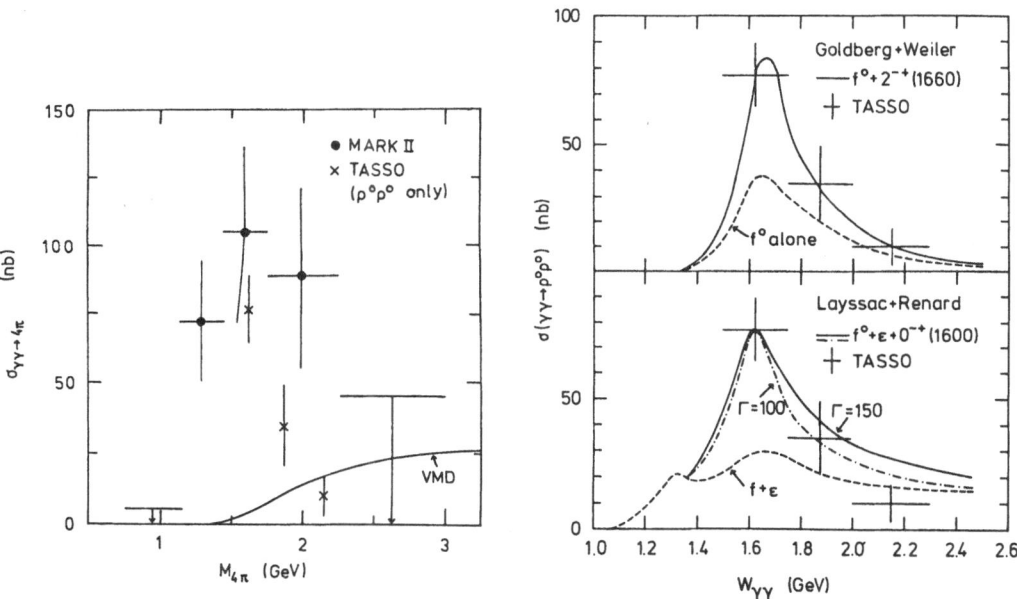

Fig.6.29.   Cross   section   for $\gamma\gamma \to \pi^+\pi^-\pi^+\pi^-$ (Mark   II)   and $\gamma\gamma \to \rho^\circ\rho^\circ$ (TASSO). The curve is explained in the text.

Fig.6.30.   Measured   cross   section   for $\gamma\gamma \to \rho^\circ\rho^\circ$ compared to theoretical models.

photoproduction and hadronic cross sections at low energies. The factorization approach suggests that the enhancement can be explained by Regge exchanges. Other authors, however, argue that it is not possible to account for the measured $\rho^\circ\rho^\circ$ production near threshold by Regge t-channel exchanges, because the $W_{\gamma\gamma}$ dependence is too steep to be accounted for by a Regge trajectory /140/.

In most papers dealing with $\rho^\circ\rho^\circ$ production by two photons the large cross section near threshold is explained by resonances /140 to 145/: either a normal quark model resonance or, more exotically, a four-quark bound state or a glueball. In Fig.6.30 two examples of such models are shown together with the measured cross section /136/. In one case the enhancement is explained by an η recurrence with $J^P=2^-$ /141/, in the other case by a glueball with $J^P=0^-$ /140/.

Another approach is pursued in /146/. The authors consider diagrams where each photon couples to a quark pair. Evaluating the overlap integrals of the quark wave functions they derive a (non-resonant) threshold enhancement for vector meson pair production.

Angular Correlation Analysis (TASSO):

The TASSO collaboration has addressed the question of whether the large $\rho^\circ\rho^\circ$ cross section is due to a resonance by performing a partial wave analysis of the reaction $\gamma\gamma\to\rho^\circ\rho^\circ$ in the four pion mass range $1.2<M_{4\pi}<2.0$ GeV /114/. If the two ρ°'s form a resonance, the spin and parity of the resonance can be determined by analysing the angular correlations in the four pion final state.

The analysed data sample corresponds to an integrated luminosity of about 40 pb^{-1}. The observed mass spectrum of events with four charged pions is shown in Fig.6.31. In this plot, four pion final states were selected by requiring that the total transverse momentum of the pions with respect to the beam axis is smaller than 0.15 GeV. It has been explained in Chap.2 that this cut selects events with no additional particles escaping detection (see Fig.2.4). The evidence for $\rho^\circ\rho^\circ$ production is demonstrated in Fig.6.32 for two different $M_{4\pi}$ mass ranges. For each event one $\pi^+\pi^-$ mass is plotted versus the other (two entries per event). The distributions show a clear enhancement in the $\rho^\circ\rho^\circ$ mass region, which is not present in the 'wrong' mass combination $\pi^+\pi^+$ and $\pi^-\pi^-$ (one entry per event). The first $M_{4\pi}$ region in this figure is below the nominal $\rho^\circ\rho^\circ$ threshold (~1.5 GeV). Although in this region the $\rho^\circ\rho^\circ$ peak cannot be reached kinematically, a strong clustering in the unlike sign mass combinations close to the kinematical boundary indicates clearly 'off-shell' $\rho^\circ\rho^\circ$ production.

In the following we define the matrix elements for $\rho^\circ\rho^\circ$ production in the spin-parity states $J^P = 0^+$, 0^-, 2^+ and 2^-, which were used in the partial wave analysis (spin-1 states can be neglected for quasi-real photons).

In the no-tag case the four pion final state is defined by 7 variables. One can choose two masses (m_{12}, m_{34}), the ρ production angle ϑ_ρ^{12} and two angles for each decaying ρ (ϑ_π^{12}, φ_π^{12}, ϑ_π^{34}, φ_π^{34}). The indices 12 and 34 refer to a pion pair as defined if one numbers the pions as follows: $\pi^+{}_1$, $\pi^-{}_2$,

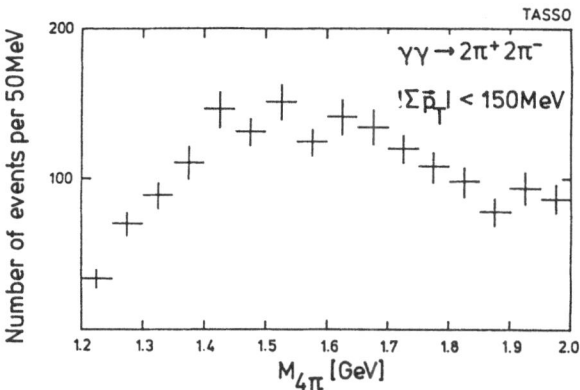

Fig.6.31. $\pi^+\pi^-\pi^+\pi^-$ mass distribution (TASSO).

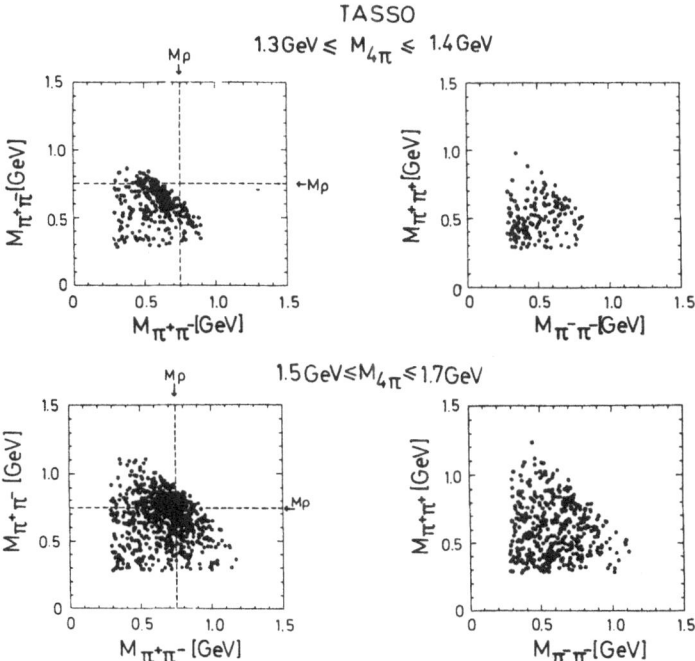

Fig.6.32. Scatter plot of $\pi\pi$ mass combinations: One $\pi\pi$ mass combination is plotted versus the other one for $\pi^+\pi^-\pi^+\pi^-$ events in two different $M_{4\pi}$ intervals; left side: $\pi^+\pi^-$ versus $\pi^+\pi^-$; right side $\pi^-\pi^-$ versus $\pi^+\pi^+$.

π^+_3, π^-_4. The rotational properties of the $\rho^\circ\rho^\circ$ system with spin-parity J^P and helicity J_z are then given by:

$$\psi^{JP,Jz} \propto \sum a_{L,Lz,S12,S34}^{JP,Jz} \, Y_L^{Lz}(\vartheta_\rho^{12},\varphi_\rho^{12}) \, Y_1^{S12}(\vartheta_\pi^{12},\varphi_\pi^{12}) \, Y_1^{S34}(\vartheta_\pi^{34},\varphi_\pi^{34}).$$
$$(6.56)$$

The sum goes over L, L_z, S_{12} and S_{34}. L is the total orbital angular momentum of the $\rho^\circ\rho^\circ$ system and L_z its projection on the z axis. S_{12} (S_{34}) is the z projection of the spin of the ρ° that decays into pions 1 and 2 (3 and 4). The angles $\vartheta_\rho{}^{12}$ and $\varphi_\rho{}^{12}$ describe the production of the ρ's in the $\gamma\gamma$ center of mass system and $\vartheta_\pi{}^{12}$, $\varphi_\pi{}^{12}$ ($\vartheta_\pi{}^{34}$, $\varphi_\pi{}^{34}$) describe the decay of the two ρ°'s in their rest system. The dependence on $\varphi_\rho{}^{12}$ is integrated out if the scattered electrons are not detected. The coefficients $a_{L,Lz,S12,S34}{}^{JP,Jz}$, as given in Table 6.5 /114/, have been evaluated by considering only the lowest orbital angular momentum L of the $\rho^\circ\rho^\circ$ system, i.e. L=0 for even parity and L=1 for odd parity. In the $\gamma\gamma$ initial state only the lowest multipole was considered. The latter restriction fixes the relative strength of the helicity components J_z for a given J^P. One finds that only J_z=0 contributes for J=0 and J^P=2$^-$; for J^P=2$^+$ the intensity of the J_z=2 component is six times larger than the J_z=0 component. This result reproduces the helicity 2 dominance for 2^{++} mesons, which is predicted to hold also under more general assumptions (see Sect.6.4.1). Note that, with the above restrictions, the angular distributions are unambiguously defined.

The matrix element for $\rho^\circ\rho^\circ$ production in a definite spin-parity state has been defined as:

$$g_{\rho\rho}{}^{JP,Jz} = \sqrt{\tfrac{1}{2}} \left(BW(m_{12})\ BW(m_{34})\ \psi^{JP,Jz}(\vartheta_\rho{}^{12},\vartheta_\pi{}^{12},\varphi_\pi{}^{12},\vartheta_\pi{}^{34},\varphi_\pi{}^{34}) \right.$$
$$\left. + BW(m_{14})\ BW(m_{32})\ \psi^{JP,Jz}(\vartheta_\rho{}^{14},\vartheta_\pi{}^{14},\varphi_\pi{}^{14},\vartheta_\pi{}^{32},\varphi_\pi{}^{32}) \right). \quad (6.57)$$

BW(m) is the ρ° Breit-Wigner amplitude. The matrix element is symmetric in the two possible $\pi^+\pi^-$ mass combinations that can form two ρ's out of the four pions. In addition a matrix element for $\rho^\circ\rho^\circ$ production according to $\rho^\circ\rho^\circ$ phase space with isotropic and uncorrelated production and decay of the ρ's was defined by setting $\psi^{JP,Jz}$=1 in (6.57).

In Fig.6.33 experimental angular distributions are compared to Monte Carlo simulations of different spin-parity states. It is clear from the plots that the negative parity states have clear and distinct signatures which are not observed in the data. Pure $\rho^\circ\rho^\circ$ phase space seems to describe the data quite well. Within the limited acceptance the distributions for J^P=0$^+$ and especially for J^P=2$^+$ are not very different from the isotropic case.

Table 6.5.   Coefficients for $\rho^\circ\rho^\circ$ angular distributions as defined in (6.56).

L	L_z	S_{12}	S_{34}	0$^+$	0$^-$	2$^+$ J_z=0	2$^+$ J_z=2	2$^-$ J_z=0
0	0	1	-1	1		1		
0	0	0	0		-1	2		
0	0	-1	1	1		1		
0	0	1	1				1	
1	1	0	-1		1			1
1	1	-1	0		-1			-1
1	0	1	-1		-1			2
1	0	-1	1		1			-2
1	-1	0	1		-1			-1
1	-1	1	0		1			1

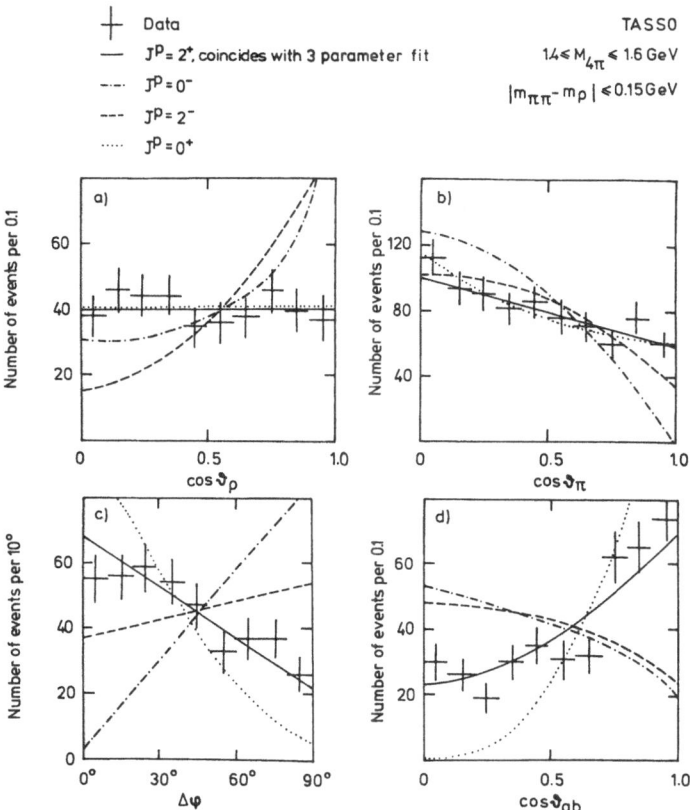

Fig.6.33. Angular distributions obtained in the analysis of $\gamma\gamma\to\rho^\circ\rho^\circ$ (TASSO): Plotted are angular distributions without acceptance corrections for $\pi^+\pi^-\pi^+\pi^-$ events in the mass range $1.4 < M_{4\pi} < 1.6$ GeV. The curves are Monte Carlo simulations for different spin-parities and include the detector effects. ("3 parameter fit" refers to the fit with isotropic $\rho^\circ\rho^\circ$ production; ϑ_ρ: ρ production angle; ϑ_π: ρ decay angle; $\Delta\varphi$: angle between the decay planes of the ρ's. ϑ_{ab}: angle between the decay directions of the ρ's. The ρ decay angles are defined in the helicity system of each ρ.)

The one-dimensional distributions cannot contain all information about the final state. Therefore the final results were obtained with a maximum likelihood fit, which included all information about correlations in angles and masses. Two types of fits have been done:

1. The $\rho^\circ\rho^\circ$ contribution was described by the four J^P states, 0^+, 0^-, 2^+, 2^-.
2. The $\rho^\circ\rho^\circ$ production was assumed to be isotropic.

In both cases possible contributions from $\gamma\gamma\to\rho^\circ\pi^+\pi^-$ and $\gamma\gamma\to\pi^+\pi^-\pi^+\pi^-$ (phase space) have been allowed for. The fits were done independently in 100 MeV intervals in $M_{4\pi}$. The cross sections determined by the fits for the four spin-parity states are plotted in Fig.6.34. Here and in the next figure the cross sections have been obtained using data for an integrated luminosity of about 80 pb^{-1} /53/ (the published analysis /114/ included ~40 pb^{-1}). The important results of the fits are:

1. The negative parity states 0$^-$ and 2$^-$ cannot be dominant in any $M_{4\pi}$ interval.
2. The $\rho^\circ\rho^\circ$ contribution is best fitted by the positive parity states 0$^+$ and 2$^+$, with a preference for 0$^+$ below ~1.7 GeV and for 2$^+$ above ~1.7 GeV.
3. The $\rho^\circ\rho^\circ$ contribution can be equally well described by isotropic $\rho^\circ\rho^\circ$ production only.

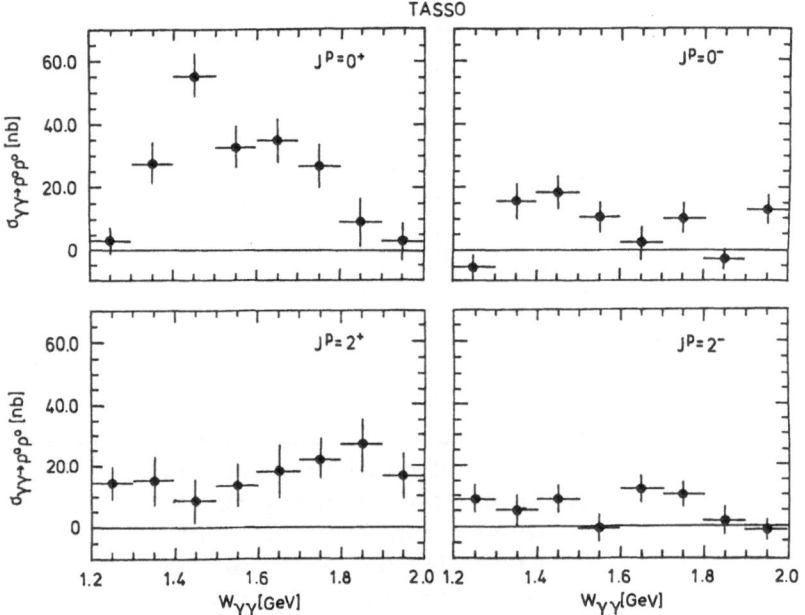

Fig.6.34.   Spin-parity decomposition of the cross section for $\gamma\gamma\rightarrow\rho^\circ\rho^\circ$ assuming contributions from   J^P=0$^+$,   J^P=0$^-$,   J^P=2$^+$ and   J^P = 2$^-$ (TASSO).

This ambiguity, namely that the spin-parity states and the isotropic $\rho^\circ\rho^\circ$ production both fit the data, could not be resolved in this analysis. It may be that the chosen matrix element does not exactly describe the physical process. In fact, indications for final state interactions were found by studying correlations in the $\pi\pi$ masses /53/. Such final state interactions are not included in the matrix elements used in the fits.

The $\rho^\circ\rho^\circ$ cross section, as obtained by fitting a $\rho^\circ\rho^\circ$ phase space contribution and contributions from $\rho^\circ\pi^+\pi^-$ and $\pi^+\pi^-\pi^+\pi^-$ phase space to the data, is shown in Fig.6.35a. Figure 6.35b shows the cross section derived

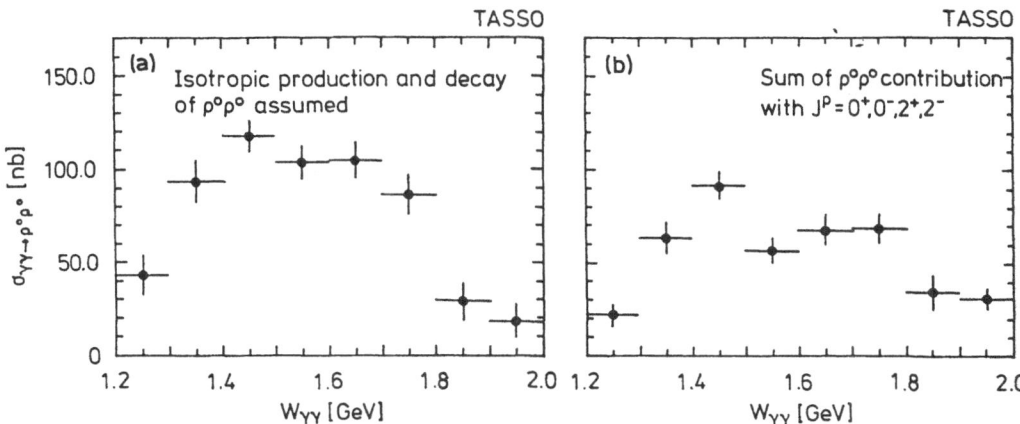

<u>Fig.6.35.</u>   Cross section for $\gamma\gamma \to \rho^\circ\rho^\circ$ (TASSO): a) Isotropic $\rho^\circ\rho^\circ$ pro-
duction and decay assumed; b) sum of the contributions from
$J^P=0^+,0^-,2^+,\ 2^-$ as obtained from the fit.

from a fit with the four spin-parity states instead of isotropic $\rho^\circ\rho^\circ$ (fit with
6 parameters). Note that the detector efficiencies are strongly dependent
on the angular distribution assumed. The differences between the two de-
terminations of the cross section reflect the systematic uncertainty of the
result, since both fits are acceptable.

Despite the differences in absolute magnitude, the fact that the cross
section stays high below the nominal $\rho^\circ\rho^\circ$ threshold is common to both
solutions. This is surprising because the available phase space decreases
drastically, implying that the matrix element responsible for $\rho^\circ\rho^\circ$ production has to increase rapidly towards lower $W_{\gamma\gamma}$.

One may ask whether known resonances could explain the observed
$\rho^\circ\rho^\circ$ cross section. From the behaviour of the matrix element one may
suspect the f(1270) of causing the threshold effect. However, using the
measured $\gamma\gamma$ width of the f (Table 6.2) and the VMD model, the f contribution to the $\rho^\circ\rho^\circ$ production was estimated to be at most ~15 nb at any $W_{\gamma\gamma}$
value /140/. From the measured partial widths $\Gamma(f \to \pi^+\pi^-\pi^+\pi^-)=5.0\pm0.8$ MeV
/147/ and $\Gamma(f \to \gamma\gamma)=2.95\pm0.3$ keV (Table 6.2) one obtains at the f mass:

$$\sigma_{peak}(\gamma\gamma \to f \to \pi^+\pi^-\pi^+\pi^-) \ = \ 15 \ \pm \ 3 \ \text{nb}.$$

It is not known what fraction of the four pion decay of the f proceeds via
$\rho^\circ\rho^\circ$.

Since the partial wave analysis indicates sizable $J^P=0^+$ contributions be-
low $M_{4\pi}\sim1.7$ GeV one should also consider $\gamma\gamma$ production of the $\varepsilon(1300)$ with
subsequent decay into $\rho^\circ\rho^\circ$. Table 6.2 contains the upper limit for the $\gamma\gamma$
coupling of the $\varepsilon(1300)$ times its branching ratio into $\pi^+\pi^-$ /44/:

$$\Gamma(\varepsilon \to \gamma\gamma)\cdot B(\varepsilon \to \pi^+\pi^-) \ < \ 1.5 \ \text{keV (95\% c.l.)} \quad \text{(for } 1.3<M_\varepsilon<1.5 \text{ GeV).}$$

Assuming in addition that one can relate $\Gamma(\varepsilon \to \gamma\gamma)$ to $\Gamma(\varepsilon \to \rho^\circ\rho^\circ)$ via VMD and
assuming that the ε couples dominantly to $\pi\pi$ (i.e. $B(\varepsilon \to \pi^+\pi^-) \sim 60\%$), one

obtains a contribution of at most 5 nb (<10% of the measured cross section) from the ε for a mass ranging from 1.3 to 1.5 GeV. Note that this limit does not hold for a scalar resonance not decaying dominantly into $\pi\pi$.

The 2^+ intensity (Fig.6.34) around 1.7 GeV could be related to the state $\Theta(1640)$ (see Sect.6.4.5), which most probably has the quantum numbers $J^{PC}=2^{++}$. The Mark II group observed a $\rho^\circ\rho^\circ$ signal in radiative ψ decays which could be due to the Θ /113/. Even if that state is a glueball, as some authors suggest, it might have a non-zero $\gamma\gamma$ width due to mixing with ordinary quark states. The TASSO group finds the following upper limit

$$\Gamma(\Theta\rightarrow\gamma\gamma)\cdot B(\Theta\rightarrow\rho^\circ\rho^\circ) < 1.2 \text{ keV} \quad (95\% \text{ c.l.}).$$

The $\iota(1440)$ is another state which is currently being discussed as a glueball candidate. This state, which most probably has $J^P=0^-$, was also found to be copiously produced in radiative ψ decays. Sizeable partial widths in both $\gamma\gamma$ and $\rho^\circ\rho^\circ$ have been predicted by some models (see discussion in Sect.6.3.6). Taking the fitted 0^- contribution the TASSO group gives the following upper limit

$$\Gamma(\iota\rightarrow\gamma\gamma)\cdot B(\iota\rightarrow\rho^\circ\rho^\circ) < 1.0 \text{ keV} \quad (95\% \text{ c.l.}).$$

This is in contradiction with the prediction of /94/.

The CELLO Analysis of $\gamma\gamma\rightarrow\pi^+\pi^-\pi^+\pi^-$:

The CELLO collaboration analysed the four charged pion final state using data from an integrated luminosity of 11.3 pb^{-1} /138/. The event selection can be briefly summarized as follows: The selected events had to have four charged tracks with net charge 0 and with momenta p$\geq$120 MeV detected in a polar angular range $|\cos\vartheta|\leq0.95$ (TASSO: $|\cos\vartheta|\leq0.84$). The total transverse momentum of the four pion system was required to be $|\sum\vec{p}_T|<120$ MeV. After these cuts 835 events with an estimated background of 9% were found in the range $1.1 \leq M_{4\pi} \leq 2.5$ GeV.

In a first step the CELLO group determined the cross section for the production of four charged pions including possible intermediate resonance states. To account for differences in the acceptance depending on whether resonances are formed or not, the acceptance was calculated as a function of the masses m_{12}, m_{34}, m_{14}, m_{32} of $\pi\pi$ combinations (the indices of the masses have been explained in the previous section). Since for a full determination of the system 7 variables are needed, additional information from the measured angular distributions was included in the acceptance calculations. The ρ production was assumed to be isotropic, as justified by the distribution in Fig.6.36a. Different decay angular distributions of the ρ's in the ρ helicity system (decay angle $\Theta_\pi{}^H$) were used for different ρ production angles. The ρ helicity system is the ρ center of mass system with the quantization axis along the direction of flight of the ρ. Figure 6.36b,c show that the decay angular distribution is flat for $|\cos\Theta_\rho|<0.8$ and is consistent with a $\sin^2\Theta_\pi{}^H$ behaviour (indicative for ρ's

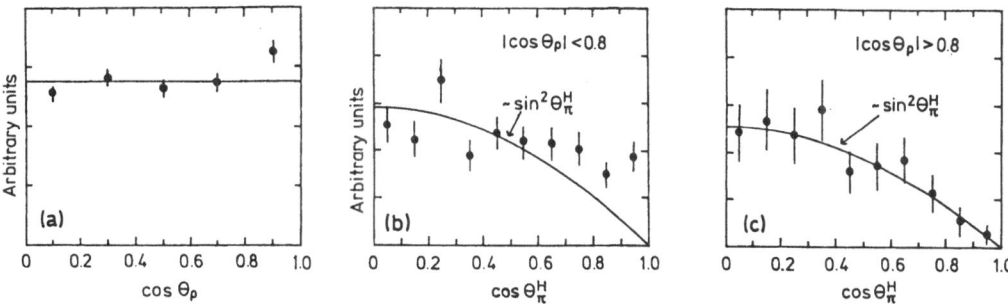

<u>Fig.6.36.</u>   Angular distributions obtained in the analysis of $\gamma\gamma\rightarrow\rho^\circ\rho^\circ$ (CELLO): The distributions include $\rho^\circ\rho^\circ$ masses between 1.3 and 2.3 GeV and are corrected for detector acceptance. "$\rho^\circ\rho^{\circ}$" is defined by cuts in the $\pi^+\pi^-$ masses (±100 MeV) around the nominal ρ mass.
a) Polar angular distribution of the produced ρ's with respect to the γ direction in the $\gamma\gamma$ center of mass system.
b) and c) Decay angular distribution of the ρ's in the ρ helicity system for two $\cos\Theta_\rho$ intervals.

with helicity ±1) for $|\cos\Theta_\rho|>0.8$. Using this angular dependence for the acceptance corrections, the four charged pion cross section in Fig.6.37a is obtained.

In a second step the CELLO group separated the three contributions $\rho^\circ\rho^\circ$, $\rho^\circ\pi^+\pi^-$ and 4π phase space in the range $1.1 \leq M_{4\pi} \leq 2.5$ GeV. For each contribution the number of events in 200 MeV wide $M_{4\pi}$ interval was deter-

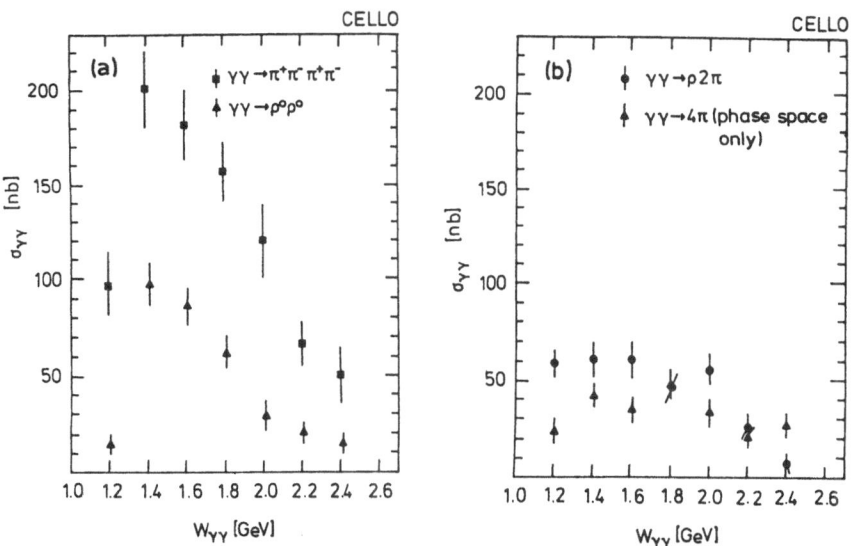

<u>Fig.6.37.</u>   Cross sections for $\gamma\gamma\rightarrow\pi^+\pi^-\pi^+\pi^-$ (CELLO): a) Cross section for $\gamma\gamma\rightarrow\pi^+\pi^-\pi^+\pi^-$ (all 4 pion final states) and for $\gamma\gamma\rightarrow\rho^\circ\rho^\circ$. b) Cross section for $\gamma\gamma\rightarrow\pi^+\pi^-\pi^+\pi^-$ (phase space only) and $\gamma\gamma\rightarrow\rho^\circ\pi^+\pi^-$.

mined by fitting the sum of the three contributions to the two density distributions $m(\pi^+\pi^-)$ versus $m(\pi^+\pi^-)$ and $m(\pi^+\pi^+)$ versus $m(\pi^-\pi^-)$.

The $\rho^\circ\rho^\circ$ cross section is shown in Fig.6.37a. The shape and magnitude of the cross section are in good agreement with the TASSO result. The $\rho^\circ\rho^\circ$ channel accounts for about half of the four charged pion cross section. The cross sections for $\rho^\circ\pi^+\pi^-$ and 4π phase space production are shown in Fig.6.37b. Note that the $\rho^\circ\pi^+\pi^-$ cross section, though of similar size, does not exhibit a threshold behaviour as dramatic as observed for $\rho^\circ\rho^\circ$, since the $\rho^\circ\pi^+\pi^-$ threshold is much lower ($\sim$1 GeV).

Summary of the Results on the Four Charged Pion Final State:

The results of the measurements of the four charged pion final state are summarized below: The large cross section for $\rho^\circ\rho^\circ$ production extends below the nominal $\rho^\circ\rho^\circ$ threshold. A sizeable $\rho^\circ\pi^+\pi^-$ cross section is observed. The partial wave analysis for $\rho^\circ\rho^\circ$ yields: Positive parity states dominate, $J^P = 0^+$ below $\sim$1.7 GeV and $J^P = 2^+$ above $\sim$1.7 GeV. There is no significant contribution from $J^P = 0^-$ and 2^-.

It appears to be excluded that well-known resonances could acount for a $\rho^\circ\rho^\circ$ enhancement with spin-parity 0^+ and 2^+. Many other explanations for the large $\rho^\circ\rho^\circ$ production have been discussed. A clue for the solution of this question is expected from measurements of the production of other vector meson pairs. Table 6.6 shows that different models can differ largely in their predictions for the relative yields of vector meson pair production.

A particularly clean test for resonance production is provided by a comparison of the isospin related channels $\rho^+\rho^-$ and $\rho^\circ\rho^\circ$. The usual $q\bar{q}$ resonances, which have isospin I=0 if they decay into $\rho^\circ\rho^\circ$, decay twice as frequently into $\rho^+\rho^-$ as into $\rho^\circ\rho^\circ$. A resonance which does not fulfil this condition must be exotic, e.g. a four-quark state.

Table 6.6.   Model predictions for the relative strength of of two-photon production of vector meson pairs: Listed are the ratios of vector meson pair production to $\rho^\circ\rho^\circ$ production for the maxima of the predicted cross sections.

Model	$\rho^\circ\rho^\circ$	$\rho^+\rho^-$	$\omega\omega$	$\rho^\circ\omega$	$\varphi\varphi$	$\rho\varphi$
VMD (γV-coupling)	1	0	1/81	1/9	4/81	2/9
VMD+Regge /139/	1	-	5.8	0.5		
Quark model /149/	1	2	1	36/25		
Quark model /146/	1	4/25*	-	-	4/25	-
Resonance (I=0)	1	2	-	0	-	0
$qq\bar{q}\bar{q}$ /145/	1	$\sim$0	$\sim$0.03	$\sim$0.6	$\sim$0.05	$\sim$0.6
$qq\bar{q}\bar{q}$ /144/	1	$\sim$0	$\sim$0.06	$\sim$0.03	$\sim$0.01	$\sim$0.1

*) It has been pointed out /150/ that this ratio should be 32/25 if one correctly accounts for the diagrams, which should contribute according to the model of /146/.

6.7.2   The Two-Photon Production of $\rho^+\rho^-$, $\rho^0\omega$, $\omega\omega$

The JADE group searched for the reaction

$$\gamma\gamma \rightarrow \rho^+\rho^- \rightarrow \pi^+\pi^-\pi^0\pi^0 \tag{6.58}$$

in a data sample collected for an integrated luminosity of 77 pb^{-1} /53/. Events were selected requiring two oppositely charged tracks with momenta p>100 MeV and four photons with shower energies E_γ>60 MeV each. Photon pairs with invariant masses between 60 and 220 MeV were considered to be π^0's. For the subsequent analysis the kinematical variables of the two photons were fitted with a constraint to the π^0 mass.

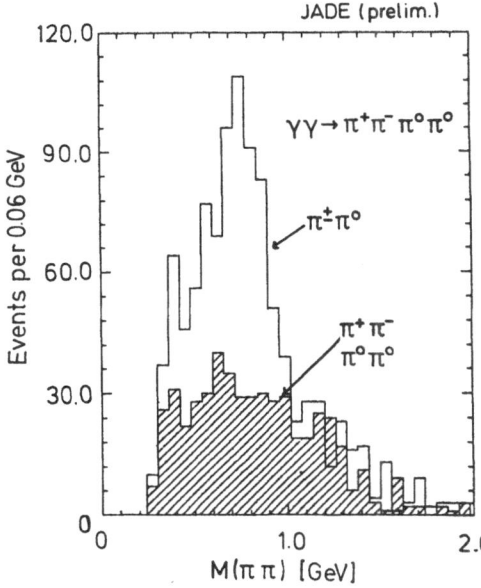

Fig.6.38.   Plot of two-pion masses in $\gamma\gamma\rightarrow\pi^+\pi^-\pi^0\pi^0$ events (JADE): The open histogram shows the charged pion pairs (4 entries per event) and the hatched histogram the neutral pion pairs (2 entries per event).

Plotting the masses of the $\pi^+\pi^0$ and $\pi^-\pi^0$ combinations (Fig.6.38) one obtains a signal at the ρ mass, which is not seen in the neutral combinations $\pi^+\pi^-$ and $\pi^0\pi^0$. The absence of a ρ^0 signal in the $\pi^+\pi^-$ combination is easily explained: Since the four pion final state with charge conjugation C=+1 has even isospin I, the presence of a ρ^0 with I=1 requires that the $\pi^0\pi^0$ combination be also in an I=1 state which is not possible for $\pi^0\pi^0$. Although there is a charged ρ signal, the two-dimensional distribution of $\pi^+\pi^0$ masses versus $\pi^-\pi^0$ masses in Fig.6.39a does not show a clear evidence for correlated $\rho^+\rho^-$ production. For comparison the corresponding plot for the neutral mass combinations is shown in Fig.6.39b.

The measured event rate is plotted in Fig.6.40 as a function of the four pion mass (open histogram). The hatched histogram is the rate restricted to the $\rho^+\rho^-$ band (0.55 < m_ρ < 0.95 GeV). The observed shift of the distribution after this cut may be just a phase space effect. The cross section

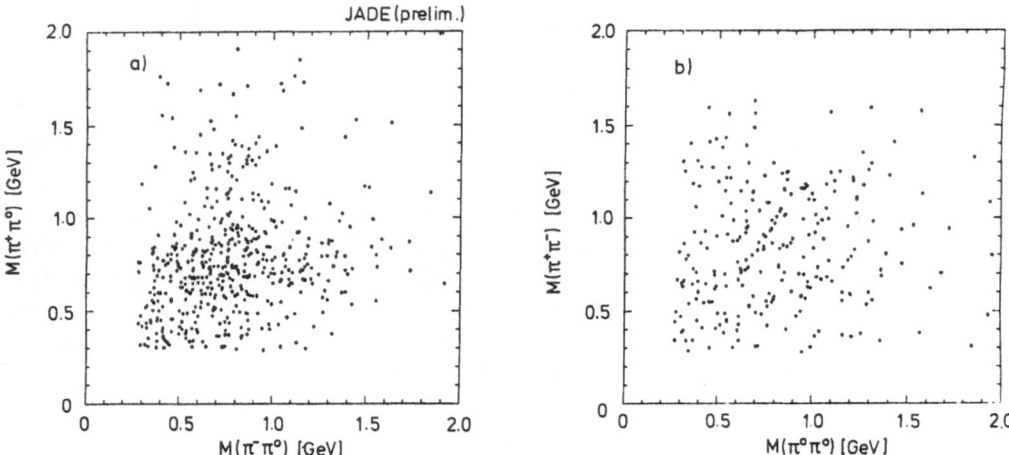

Fig.6.39.   Scatter plot of $\pi\pi$ masses in $\gamma\gamma\rightarrow\pi^{+}\pi^{-}\pi^{0}\pi^{0}$ events (JADE): a) charged pion pairs, 2 entries per event; b) neutral pion pairs, 1 entry per event.

derived from the events in the $\rho^{+}\rho^{-}$ band is plotted in Fig.6.41. Since the $\rho^{+}\rho^{-}$ rate has not been explicitly determined, this cross section has to be considered as an upper bound for $\sigma(\gamma\gamma\rightarrow\rho^{+}\rho^{-})$. The errors in the plot are statistical only; the systematic errors are about 30%.

The JADE group also analysed the channels /81/

$$\gamma\gamma \rightarrow \rho^{0}\omega \rightarrow \pi^{+}\pi^{-}\pi^{+}\pi^{-}\pi^{0} \tag{6.59}$$

$$\gamma\gamma \rightarrow \omega\omega \rightarrow \pi^{+}\pi^{-}\pi^{0}\pi^{+}\pi^{-}\pi^{0}. \tag{6.60}$$

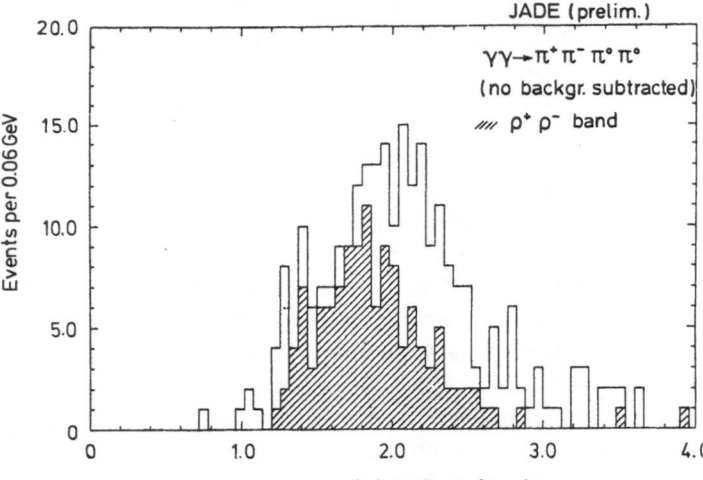

Fig.6.40.   Four pion mass spectrum of the reaction $\gamma\gamma\rightarrow\pi^{+}\pi^{-}\pi^{0}\pi^{0}$ (JADE). The hatched histogram is the same spectrum with mass combinations in the $\rho^{+}\rho^{-}$ band.

In both cases no signal was observed and the upper limits in Fig.6.42 were derived. For comparison the cross section for $\gamma\gamma \to \rho^\circ\rho^\circ$ as measured by TASSO is also shown in this plot.

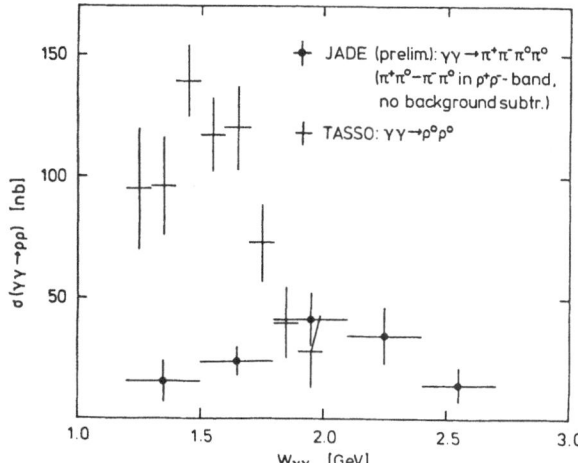

Fig.6.41.   Cross section for $\gamma\gamma \to \rho^+\rho^-$ (JADE, preliminary): The cross section is obtained from the events in the hatched histogram of the previous figure. Since no background has been subtracted the cross section has to be taken as an upper bound. Also shown is the cross section for $\gamma\gamma \to \rho^\circ\rho^\circ$ from TASSO /114/.

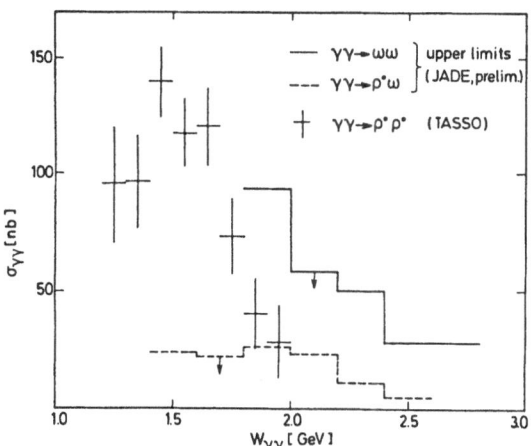

Fig.6.42.   Upper limits for $\gamma\gamma \to \rho^\circ\omega$ and $\gamma\gamma \to \omega\omega$ (JADE, preliminary). Also shown is the cross section for $\gamma\gamma \to \rho^\circ\rho^\circ$ from TASSO /114/.

6.7.3   Discussion of the Results on $\gamma\gamma$ Production of Vector Meson Pairs

The upper limits given by the JADE group for the $\gamma\gamma$ production of $\rho^+\rho^-$, $\rho^\circ\omega$ and $\omega\omega$ near threshold impose rather stringent constraints on the models proposed to explain the $\rho^\circ\rho^\circ$ threshold enhancement (see Table 6.6). In particular, since the $\rho^+\rho^-$ production does not show this enhancement, we conclude that the $\rho^\circ\rho^\circ$ enhancement cannot be due to a single resonance with well defined isospin. The relative rates expected for the decay of a resonance with isospin I (I = 0, 2) into pairs of neutral and charged ρ's are:

I	$\rho^+\rho^-$	$\rho^\circ\rho^\circ$
0	2	1
2	1	2

The experimental result for $\rho^+\rho^-$ production may also be in contradiction to the model of /146/ (see footnote in Table 6.6). The model of /139/ predicts particularly large $\omega\omega$ production, which is ruled out by the measured upper limit for this channel.

What possible explanations for the $\rho^\circ\rho^\circ$ enhancement are now remaining? The naive model, in which the cross sections for vector meson pair production are proportional to the VMD coupling constants (first line in Table 6.6), is not yet ruled out. Such a model implies vector meson scattering without quantum number exchange, which is expected to describe the high energy $\gamma\gamma$ total cross section. However, there is probably no good reason for this model to work at low energies. The four-quark model /144, 145/ explains quite naturally the suppression of $\rho^+\rho^-$ relative to $\rho^\circ\rho^\circ$ by an interference effect between isoscalar and isotensor states. It is also not in disagreement with the upper limits of $\rho^\circ\omega$ and $\omega\omega$ production. In the following we discuss some of the features of this model which are relevant to two-photon production of vector meson pairs.

The four-quark states are ordered in multiplets. The lowest multiplet is a $J^{PC} = 0^{++}$ nonet with masses around 1 GeV. It has been suggested that the scalar states $\varepsilon(700)$, $S^*(975)$ and $\delta(980)$ belong to this multiplet /67, 151/. These states couple dominantly to a pair of pseudoscalar mesons (PP) and the $\gamma\gamma$ width should be suppressed (up to now in agreement with the experiments). According to VMD we expect the largest couplings to $\gamma\gamma$ for the multiplets which have a dominant coupling to pairs of vector mesons (VV). Table 6.7 contains a list of some of the expected multiplets together with their average masses and the relative amplitudes for the couplings to pseudoscalar and vector meson pairs ('recoupling coefficients'). In general, a four-quark state should be broad because it can easily decay into pairs of $q\bar{q}$ mesons ('superallowed' decays). An exception could be, for example, the 0^{++} state at about 1.45 GeV (Table 6.7), which couples dominantly to $\rho^\circ\rho^\circ$. This state is predicted to have a relatively narrow total width (i.e. typical hadronic width) because it lies below the threshold for its superallowed decay into $\rho^\circ\rho^\circ$ /151/. Above threshold the widths become

Table 6.7. The recoupling coefficients of the S wave four-quark states which decay mainly into two vector mesons: PP and VV denote the amplitudes for decays into pairs of pseudoscalar and vector quark-antiquark states, respectively.

Multiplet	$J^{PC}(I)$	Mass [GeV]	PP	VV
9*	$0^{++}(0)$	1.45	−0.177	0.644
36*	$0^{++}(0,2)$	1.80	0.041	0.743
9	$2^{++}(0)$	1.65	0	0.816
36	$2^{++}(0,2)$	1.65	0	0.577

gradually larger, so that around 2 GeV the states coupling to $\rho\rho$ are expected to be too broad to be seen as resonance enhancements.

The four-quark states in Table 6.7 have all positive parity, which is also the preferred assignment for the observed $\rho^0\rho^0$ production. Some of the multiplets appear both with isospin I=0 and I=2. The suppression of $\rho^+\rho^-$ production is explained in the four-quark model by a destructive interference between two different isospin states. In the framework of this model, we have to conclude that the 0^{++} state at 1.45 GeV with I=0 cannot contribute significantly, because for this state $\rho^+\rho^-$ production would be twice as large as $\rho^0\rho^0$ production. Therefore, the four-quark model also has problems to explain the large $\rho^0\rho^0$ cross section below ~1.5 GeV (although it predicts an enhancement closely above threshold). We conclude that the four-quark model explains some of the observed features of two-photon production of vector meson pairs but certainly not all, at least not without adjusting model parameters.

More experimental information is probably needed to find the correct model for the production of vector meson pairs near threshold. For a further sophistication of the experimental $\rho^0\rho^0$ analyses it would be useful, if theorists provide complete matrix elements as a function of $W_{\gamma\gamma}$. Note that none of the models has yet explained the measured $\rho^0\rho^0$ cross section below ~1.5 GeV. The information on the production of other vector meson pairs is still very limited. As more luminosity is collected, the upper limits on $\rho^+\rho^-$, $\rho^0\omega$ and $\omega\omega$ may turn into cross section measurements. The authors of /145/ emphasize that the measurement of $\rho\varphi$ production provides a crucial test of the four-quark model. The cross section is predicted to be about 60 nb near threshold. It will be difficult to detect φ's at threshold via the decay $\varphi\to K^+K^-$ because of the low Q value of this decay. Higher sensitivity may be reached using the decay mode $\varphi\to\pi^+\pi^-\pi^0$ (branching ratio 14.8%).

6.8   Observation of a Narrow Structure at 2.1 GeV in $\gamma\gamma \to \pi^+\pi^-\pi^+\pi^-$

The TASSO collaboration observed a narrow structure around 2.1 GeV in the four charged pion mass spectrum (Fig.6.43) /152, 53/. The data correspond to about 80 pb^{-1}. The event selection was the same as for the $\rho^\circ\rho^\circ$ analysis (Sect.6.7.1) except for a more stringent cut in the transverse momentum of the four pion system ($|\sum\vec{p}_T|$ <0.07 GeV).

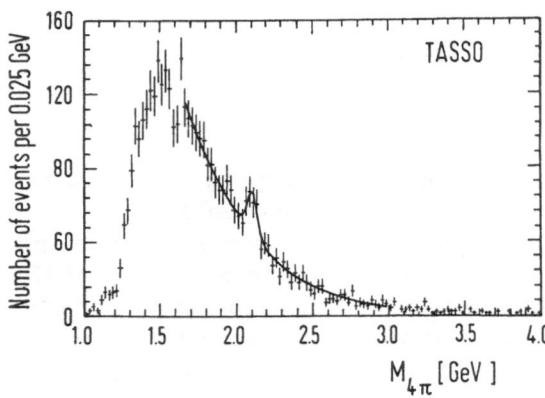

Fig.6.43.   Mass spectrum of four charged pions (TASSO).

The statistical significance of the structure was determined by fitting a smooth curve to the mass spectrum between 1.65 and 3.0 GeV including the signal region. For masses between 2.05 and 2.15 GeV the observed event rate is 4.3 standard deviations larger than expected from this fit. Including a Breit-Wigner function in addition to the smooth curve, the fit gives a good description of the data and yields for the Breit-Wigner parameters, mass and width, and for the number of events in the peak:

$$M=2.103\pm0.01 \text{ GeV}, \quad \Gamma=0.030\pm0.034 \text{ GeV}, \quad N=125.6\pm46.$$

In this fit the 4 pion mass resolution, which is about 60 MeV (FWHM) near 2 GeV, has been taken into account.

If the structure is indeed a resonance, its $\gamma\gamma$ coupling is:

$$\Gamma_{\gamma\gamma}\cdot(2J+1)\cdot B(4\pi^\pm) = 1.25 \pm 0.5 \text{ (stat.)} \pm 0.5 \text{ (syst.) keV}.$$

J is the spin of the resonance and $B(4\pi^\pm)$ its decay branching ratio into four charged pions. Upper limits for the branching ratios into $\rho^\circ\rho^\circ$ and $K_0\bar{K}_0$ are found to be

$$B(\rho^\circ\rho^\circ)/B(4\pi^\pm) < 0.6 \text{ (95\% c.l.)},$$
$$B(K_0\bar{K}_0)/B(4\pi^\pm) < 0.06 \text{ (95\% c.l.)}.$$

The signal does not seem to be correlated with the large $\rho^\circ\rho^\circ$ enhancement observed between threshold and $M_{4\pi} \sim 2$ GeV. Removing events which

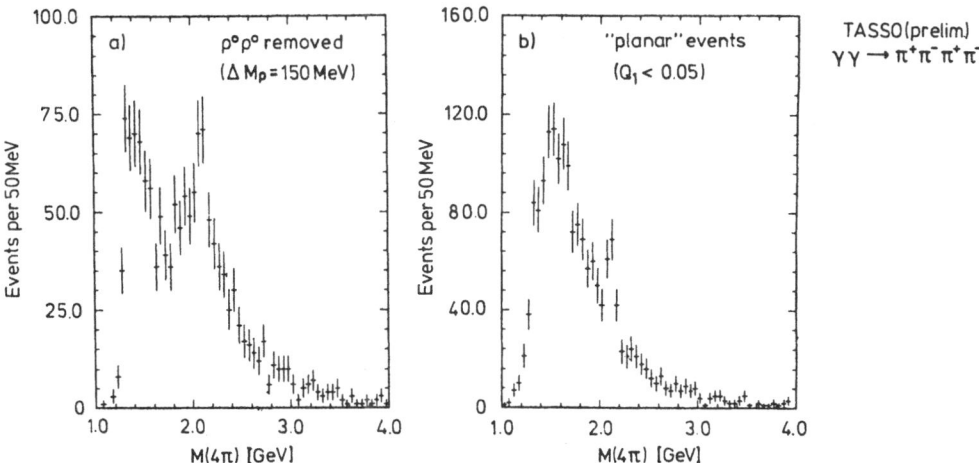

<u>Fig.6.44.</u>   Four charged pion mass spectrum with additional cuts: a)
$\rho^{\circ}\rho^{\circ}$ events removed; b) events with $Q_1 < 0.05$.

have a $\pi^+\pi^-$ - $\pi^+\pi^-$ combination in the $\rho^{\circ}\rho^{\circ}$ band (cut in the ρ° mass: ±150
MeV) the $M_{4\pi}$ distribution shown in Fig.6.44a was obtained. Although the
signal appears more pronounced in this plot, a quantitative analysis is
more difficult because the effect of the applied cut on the background introduces uncertainties. The signal to background ratio is improved if one
plots only events which appear planar in the laboratory system (Fig.6.44b).
"Planar" is defined by $Q_1 < 0.05$, where Q_1 is the smallest eigenvalue of the
sphericity tensor. Monte Carlo studies showed that the latter cut also en-
hances 4π phase space over $\rho^{\circ}\rho^{\circ}$ final states. The effects of these
kinematical cuts have to be further studied.

Should the observed narrow structure be confirmed as a resonance, it
cannot easily be associated with a known state having similar resonance
parameters. The nearest candidate would be the h(2040), but it does not
seem to have the same resonance parameters. Note however, that the authors of /97/ predicted a relatively large $\gamma\gamma$ width for the h(2040), which
they called in their paper the f recurrence f*.

7. The Total Cross Section for Two-Photon Production of Hadrons

In the next three chapters we will discuss hadron production by two photons in different kinematical regions:

- hadron production by two photons, both having small Q^2 (this chapter);
- the processes involving hadrons with large transverse momenta (Chap.8) and
- deep inelastic scattering of one highly virtual photon off a quasi-real photon (Chap.9).

7.1 Theoretical Predictions

Following the vector meson dominance model, the scattering of two quasi-real photons is expected to proceed dominantly like hadron–hadron scattering (Fig.7.1). With the probability $\alpha\pi/\gamma_V^2$ for a photon to turn into a vector meson V, one gets the relation:

$$\sigma(\gamma\gamma\to\text{hadrons}) = \sum_{V,V'} (\alpha\pi/\gamma_V^2)\cdot(\alpha\pi/\gamma_{V'}^2)\ \sigma(VV'\to\text{hadrons}) \tag{7.01}$$

$(\alpha\pi/\gamma_V^2 \approx 2.80\cdot10^{-3}, \quad 0.30\cdot10^{-3}, \quad 0.36\cdot10^{-3}$ for ρ, ω and φ, respectively).

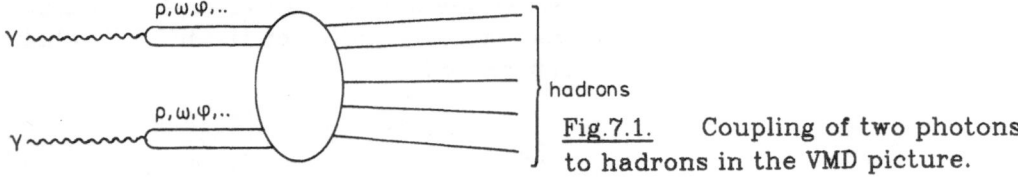

Fig.7.1.   Coupling of two photons to hadrons in the VMD picture.

In this model events from two-photon collisions should have characteristics similar to those of events from hadron-hadron scattering. That means, two-photon hadron production should be dominated by peripheral processes leading to limited transverse momenta of the hadrons with respect to the incoming beams (for low Q^2 the direction of the γ's is about the same as the beam direction). The direct, pointlike coupling of the photon to the constituents of matter is expected to yield a small contribution to the total cross section, though it may well be dominant in kinematical regions involving high momentum transfers (see next two chapters).

An estimate of the total cross section for hadron production by two real photons is given in /153/. The $\gamma\gamma$ total cross section $\sigma(\gamma\gamma\to\text{hadrons})$ is related to nucleon-nucleon (NN) scattering and γ-nucleon (γN) scattering using the Regge model together with the assumption of factorizability of the trajectories. Within the conventional Regge pole model we expect contributions from the Regge trajectories with a $W_{\gamma\gamma}$ dependence $W_{\gamma\gamma}^{2\alpha-2}$. Thus the pomeron ($\alpha=1$) trajectory contributes a constant term and the leading non-pomeron trajectories, the f and A_2 ($\alpha=\frac{1}{2}$), contribute a term $\sim 1/W_{\gamma\gamma}$:

$$\sigma(\gamma\gamma\to\text{hadrons}) = \sigma_0 + \sigma_1/W_{\gamma\gamma}. \tag{7.02}$$

Contributions of lower lying trajectories like the f', which would lead to terms with higher powers of $1/W_{\gamma\gamma}$, have been estimated to be small. A quantitative estimate for the pomeron contribution is obtained from the measured high energy behaviour of the γN and NN total cross sections:

$$\sigma_0 = \frac{[\sigma_{TOT}(\gamma N, s\to\infty)]^2}{\sigma_{TOT}(NN, s\to\infty)} \approx 240 \text{ nb.} \tag{7.03}$$

The $1/W_{\gamma\gamma}$ term (f, A_2 trajectories) has been estimated in /153/ from the corresponding terms in $p\bar{p}$, $\bar{p}n$, γp and γn scattering employing exchange degeneracy of the ω, ρ and f, A_2 trajectories. The SU(3) relation of the $\gamma\gamma$ couplings of f and A_2 in (6.29) has been used as an additional constraint. The result is:

$$\sigma_1 \approx 270 \text{ nb·GeV.} \tag{7.04}$$

This term can be related via duality to the s-channel resonances, so that the resonance contribution to the cross section in the resonance region should be given by $\sigma_1/W_{\gamma\gamma}$ on the average. In Fig.7.2 the resonances which have been measured up to now are schematically represented by rectangular boxes with areas equal to those under the corresponding resonance curves. For not too low $W_{\gamma\gamma}$ the resonances seem to saturate the curve for the $1/W_{\gamma\gamma}$ term[5]. In the same plot the $\pi^+\pi^-$ continuum as estimated by the Born approximation up to 1.5 GeV and the measured four pion cross section are added on top of the resonances (the K^+K^- continuum is much smaller). The curve for four pion production includes

---

[5]   This observation is in contrast to the conclusion drawn in /122/, that the resonance contribution is too large to be accounted for by the Regge estimate. Such a conclusion depends very much on the assumed $\gamma\gamma$ width of the scalar mesons and on the lower boundary of the cross section integral (equ. 25 in /122/).

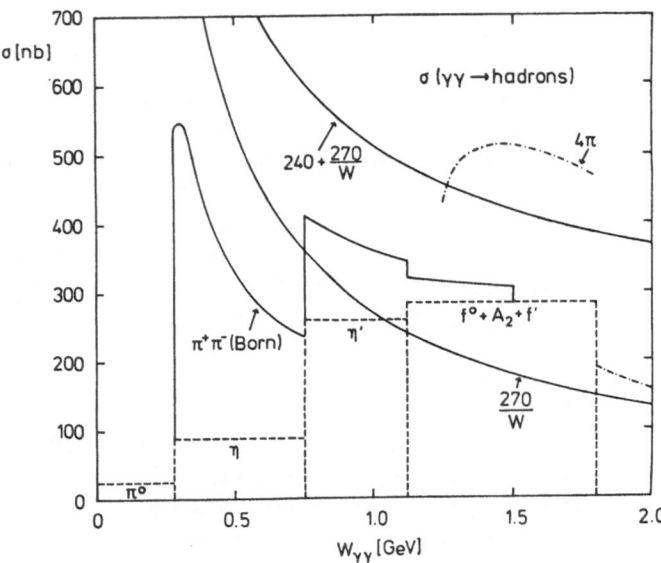

<u>Fig.7.2.</u>   Sum of measured $\gamma\gamma$ cross sections compared to predictions from Regge model calculations /153/ : The sum includes resonance production, the Born cross section for $\gamma\gamma\to\pi^+\pi^-$ and 4 pion production, $\gamma\gamma\to\pi^+\pi^-\pi^+\pi^-$ and $\gamma\gamma\to\pi^+\pi^-\pi^0\pi^0$ (restricted to the $\rho^+\rho^-$ band).

the cross section for $\gamma\gamma\to\pi^+\pi^-\pi^+\pi^-$ taken from Fig.6.37 (CELLO) and the cross section for $\gamma\gamma\to\pi^+\pi^-\pi^0\pi^0$ (restricted to the $\rho^+\rho^-$ band) from Fig.6.41 (JADE). It is interesting that above ~1 GeV not much room is left between the sum of the already measured cross sections and the predicted cross section $\sigma(\gamma\gamma\to\text{hadrons}) \approx 240$ nb $+ 270$ nb$\cdot$GeV/$W_{\gamma\gamma}$.

It has been suggested in the literature that in addition to the Regge contributions there may be non-Regge terms arising from the pointlike coupling of the photon. The motivation comes from the possible presence of fixed poles in the absorptive part of the elastic $\gamma\gamma$ scattering amplitude /154/. According to /124/ an estimate for this contribution can be obtained from the quark loop box diagram (Fig.7.3):

$$\sigma_{\gamma\gamma}^{\text{BOX}} \approx (4\pi\alpha^2/W_{\gamma\gamma}^2)\sum_q e_q^4\cdot\ln(W_{\gamma\gamma}^2/m_q^2) \tag{7.05}$$

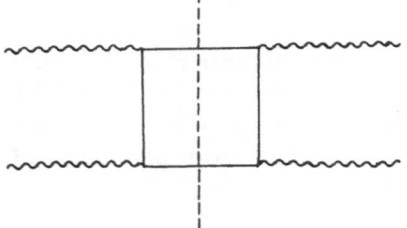

<u>Fig.7.3.</u>   Box diagram for $\gamma\gamma\to q\bar{q}$.

(e_q is the charge and m_q the mass of the quark q; the sum goes over all flavours and colours). The box diagram would thus add a term $\sigma_2/W_{\gamma\gamma}^2$ in (7.02). It has been pointed out /155/ that the box diagram of a bare-fermion loop is not the only possible source for a $1/W_{\gamma\gamma}^2$ term. The simultaneous consideration of all diagrams with gluon and fermion bubble corrections to the bare fermion loop could lead in the limit of large $W_{\gamma\gamma}$ to both Regge and fixed pole terms with a $1/W_{\gamma\gamma}^2$ dependence. The calculation of these graphs at lower energies is a problem of non-perturbative QCD and thus hard to solve with today's techniques.

In /124/ it was argued that an additional $1/W_{\gamma\gamma}^2$ term is needed to account for the resonance contributions, which has been estimated using the sum rule (6.44). With the measurements now available these estimates turn out to be too large (see also discussion in /89/). Assuming narrow resonance saturation of the sum rule as given in (6.45), the authors of /124/ find a $\gamma\gamma$ width of the f of ~9 keV which has to be compared to the measured value of 2.95±0.3 keV. Such a large width is prompted by the assumption that the $\gamma\gamma$ coupling constant of the scalar ε meson is about the same as for the π°. We have shown in Sect.6.5 that the sum rule is only consistent with the experimental results if the $\gamma\gamma$ widths of the scalars are suppressed.

7.2 The Experimental Method to Determine $\sigma(\gamma\gamma \rightarrow$ Hadrons)

Measurements of the $\gamma\gamma$ total cross section have a major disadvantage compared to the measurements of the one-photon annihilation cross section: The center of mass energy, $W_{\gamma\gamma}$, of an event is not known a priori because the photons provide a continuous spectrum of $W_{\gamma\gamma}$ values. Hence $W_{\gamma\gamma}$ has to be measured. In a double-tag experiment $W_{\gamma\gamma}$ is in principle determined by the measured kinematics of the scattered leptons. But it is difficult to obtain sufficient resolution in the tagging devices. For typical forward detectors at PETRA and PEP the resolution is not good enough for measuring the total cross section in the resonance region. In the single-tag or no-tag case, $W_{\gamma\gamma}$ has to be reconstructed from the measured hadrons. Because of the imperfect detection the measured $W_{\gamma\gamma}$ (W_{vis}) will in general be smaller than the true $W_{\gamma\gamma}$. The determination of the $W_{\gamma\gamma}$ dependence of the cross section therefore requires the use of a model, which realistically describes the hadronic final states.

7.2.1 The Double-Tag Approach

In a double-tag experiment, where both scattered electrons are detected, the $\gamma\gamma$ kinematics can be completely reconstructed from the measured lepton kinematics. With the assumption that the QED processes can be reliably subtracted (we have no reason to believe that this is not the case), this method allows one to determine the total cross section

$\sigma(\gamma\gamma\to\text{hadrons})$ in a completely unbiased and model independent way. If $0°$ tagging is possible, one has also the advantage that the real photon cross section can be measured without large extrapolations to $Q^2=0$. In principle all cross section terms (or structure functions) in (2.14) can be determined in a double-tag experiment. In practice that will be difficult to achieve for all terms, but e.g. the linear polarization asymmetry, $\tau_{TT}=\sigma_{\parallel}-\sigma_{\perp}$, may be easily obtained from the angles between the scattering planes of the electrons. Such a measurement would allow a test of the sum rule derived in /156/, which connects the asymmetry with the sum of the fourth power of the quark charges. For quasi-real photons the sum rule reads:

$$\int_0^{\infty} dW_{\gamma\gamma}^2 \ (\sigma_{\parallel} - \sigma_{\perp}) = -(\alpha^2/4\pi) \sum_q e_q^4. \tag{7.06}$$

The actual realization of a double-tag experiment encounters substantial problems. The loss in counting rate may be acceptable, if one gains an unbiased data sample in exchange. A more severe problem is the high background in the forward direction at all e^+e^- machines, which means that one cannot use e^+ and e^- detection alone. At least one more particle (usually charged) has to be required at a large enough angle and large enough transverse momentum to be detected in the central detector. Thus this measurement is no longer totally unbiased. The other major difficulty is to achieve a sufficient resolution for $W_{\gamma\gamma}$ with the tagging devices, especially at small $W_{\gamma\gamma}$. $W_{\gamma\gamma}$ is essentially given by the energy loss of the scattered leptons:

$$W_{\gamma\gamma}^2 \approx 4(E-E_1')(E-E_2'). \tag{7.07}$$

Since the energy loss tends to be small compared to the beam energy, it is difficult to obtain a good resolution. At PEP/PETRA energies the $W_{\gamma\gamma}$ resolution is about 1 GeV with lead-scintillator sandwich shower counters in the tagging devices. Using NaI crystals in the forward detector, as in the PEP-9 experiment, one can achieve a $W_{\gamma\gamma}$ resolution of about 150 MeV /157/. The BGO (= bismuth germanium oxide) tagging device proposed for the Argus detector at DORIS II should achieve a similar resolution. Lower beam energies improve the $W_{\gamma\gamma}$ resolution in double-tag experiments. We will report below on an experiment done at SPEAR at beam energies of about 3 GeV.

7.2.2   The Single-Tag Approach

Measurements of the total cross section using the single-tag method have been carried out by the PLUTO and TASSO collaborations /158, 159, 160/. Numbers quoted in the following for acceptances, resolutions etc. are typical for these experiments.

If only one scattered lepton is detected (single tag), $W_{\gamma\gamma}$ cannot be determined from the lepton kinematics. In this case $W_{\gamma\gamma}$ has to be reconstructed from the final state hadrons, which should therefore be detected as completely as possible. This requirement is at variance with the incomplete acceptance of the detectors. For one-photon annihilation events this is usually not a severe problem, but for $\gamma\gamma$ events the situation is different: due to the boost of the $\gamma\gamma$ system, the particles are preferentially produced in the forward and backward directions i.e. in the regions where the detectors have acceptance holes. The typical event detection efficiencies for one-photon events are about 80%, but for two-photon events only a few percent. For peripherally produced events, with jet-like particle distributions in the forward and backward direction, this problem becomes even more severe. In addition, the detection of neutral particles is particularly insufficient in some detectors. The reconstructed $W_{\gamma\gamma}$ (which is referred to as W_{vis}) is therefore on average smaller than the true $W_{\gamma\gamma}$, by about 25% to 60%. To obtain the $W_{\gamma\gamma}$ dependence the W_{vis} distribution has to be unfolded using a Monte Carlo simulation of the detector acceptance. Such a program needs as input a model for the two-photon cross section with the $W_{\gamma\gamma}$ dependence and the detailed structure of the hadronic final state. The parameters of such models have to be determined by the experiment unless they can be fixed by reasonable assumptions derived theoretically or from other measurements. In the single-tag case, with one quasi-real photon (as can be guaranteed by 'anti-tagging'), the formula for the $\gamma\gamma$ cross section in (2.32) averaged over $\tilde\varphi$ can be used, which depends on $W_{\gamma\gamma}$, Q_1^2 and the polarization parameter ε:

$$\sigma(W_{\gamma\gamma},Q_1^2,Q_2^2{\approx}0) = \sigma_{TT}(W_{\gamma\gamma},Q_1^2,Q_2^2{\approx}0) + \varepsilon\sigma_{LT}(W_{\gamma\gamma},Q_1^2,Q_2^2{\approx}0) . \qquad (7.08)$$

The longitudinal component contributes only for the virtual photon. Terms depending on the angle $\tilde\varphi$ between the lepton scattering planes average out to zero since the second lepton is not detected. The mass of the virtual photon, Q_1^2, is essentially determined by the tagging angle with typical resolutions of about 10% at $Q_1^2{=}0.3$ GeV^2.

In the experiments done so far, σ_{TT} and σ_{LT} have not been separated. For small tagging angles (typically 20 - 100 mrad) the polarization parameter ε of the virtual photon as given in (2.31) is approximately:

$$\varepsilon \approx \frac{2E_1'/E}{1+(E_1'/E)^2} . \qquad (7.09)$$

For a typical cut at $E_1'/E \approx 40\%$, ε varies between about 0.7 and 1.0. If σ_{TT} and σ_{LT} are not separated , the extrapolation to the cross section for real unpolarized photons, $\sigma(W_{\gamma\gamma},0,0)$ is not straightforward. We will come back to this point in the next section.

7.2.3   The Model for Hadron Production by Two Photons

The model, which is necessary to compute the detector acceptance, has to have the following features:

1. a parametrization of the $W_{\gamma\gamma}$ dependence of the total cross section;
2. a parametrization of the Q^2 dependence (to allow for the extrapolation to real photons);
3. a description of the structure of the final state.

For the $W_{\gamma\gamma}$ dependence the considerations in Sect.7.1 suggest the following form:

$$\sigma = A + B/W_{\gamma\gamma} + C/W_{\gamma\gamma}^2. \tag{7.10}$$

For the extrapolation to $Q^2=0$ a ρ pole form factor was used in /158, 160/. Such an ansatz neglects contributions from other vector mesons. Furthermore, using the same Q^2 dependence for σ_{LT} as for σ_{TT} cannot be correct because σ_{LT} has to vanish as Q^2 goes to zero (see (2.17)). The authors of /161/ pointed out that the assumption $\sigma_{LT} \ll \sigma_{TT}$ is not justified (at the Q^2 values of the TASSO and PLUTO experiments). Using the generalized vector meson dominance model (GVDM) and ep-scattering results they obtain a Q^2 dependence which also accounts for the $Q^2\rightarrow0$ limit of the longitudinal component:

$$F_{GVDM}(Q^2) = \sum_{V=\rho,\omega,\varphi} r_V \frac{1+Q^2/4m_V^2}{(1+Q^2/m_V^2)^2} + \frac{0.22}{1+Q^2/m_0^2} \tag{7.11}$$

with $r_\rho = 0.65$, $r_\omega = 0.08$, $r_\varphi = 0.05$, $m_0 = 1.4$ GeV.
Using this expression the Q^2 dependence of the cross section in (7.08) is given by (for $\varepsilon\approx1$):

$$\sigma(W_{\gamma\gamma},Q_1^2,Q_2^2\approx0) = \sigma_{TT}(W_{\gamma\gamma},Q_1^2,Q_2^2\approx0) + \varepsilon\sigma_{LT}(W_{\gamma\gamma},Q_1^2,Q_2^2\approx0) =$$
$$= \sigma(W_{\gamma\gamma},0,0)\cdot F_{GVDM}(Q_1^2). \tag{7.12}$$

For small Q^2, the Q^2 behaviour of σ_{LT} is accounted for by the term $Q^2/4m_V^2$ in (7.11). Contributions of higher mass resonances and of the continuum are approximated by the term $\sim 1/(1+Q^2/m_0^2)$. F_{GVDM} falls considerably slower with increasing Q^2 than the simple ρ pole form factor. In Fig.7.4 the ratio of both form factors is plotted versus Q^2. The difference of about 30% at $Q^2=0.3$ GeV^2 means that the extrapolation from this Q^2 to $Q^2=0$ would give a 30% difference in the cross section, depending on the form factor chosen. Another complication should be mentioned: if the $\gamma\gamma$ cross section has a contribution $\sim1/W_{\gamma\gamma}^2$ and if this contribution is due to the pointlike coupling of the photon to quarks, it would have a much weaker Q^2 dependence.

The most difficult task is the construction of a realistic model for the hadronic final state. The ingredients of such models, as used in the PLUTO and TASSO analyses, are summarized in the following.

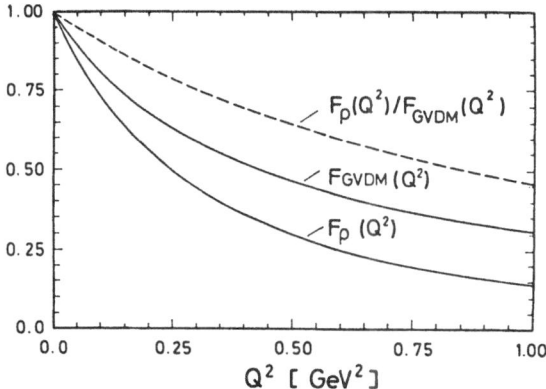

<u>Fig.7.4.</u>   Predicted Q^2 dependence of the cross section according to ρ pole dominance and the GVDM. The dashed curve is the ratio of both.

Since at small Q^2 the dominant contribution to the cross section comes from peripheral scattering, a <u>multiparticle phase space</u> with <u>limited transverse momenta</u> of the final state hadrons was used as a basis for the model. Only pions were generated, other particle species (K, p, ...) and resonance production and decay (ρ, K^0s, ...) were neglected. At this point the models can certainly be improved as experimental information about the actual particle composition in $\gamma\gamma$ reactions becomes available. The limitation of the transverse momenta with respect to the $\gamma\gamma$ direction (p_T) can be introduced into the model by a matrix element which contains for each particle i in an event a factor

$$\sim \exp(-\alpha \cdot p_{Ti}) \quad \text{or} \quad \sim \exp(-\beta \cdot p_{Ti}{}^2) \quad \text{or} \quad \sim \exp(-\gamma \cdot \sqrt{m^2 + p_{Ti}{}^2}).$$

In two-photon reactions the mass parameter in the last expression may be chosen to be the ρ mass /12/. The exponential behaviour of the p_T distribution has a strong impact on the detector efficiency because the particles are produced with the highest probability in kinematical regions where the detectors have low acceptances (at low p_T). That is especially dangerous because the p_T slope is strongly correlated with the multiplicity. Since the event generators /162/ have to account for energy conservation, high average multiplicities result in an effective p_T slope which is steeper than expected from the slope parameter. In addition, resonance formation will also change the observed p_T distribution.

For the <u>average charged multiplicity</u> an ansatz can be used which has been determined from e^+e^- annihilation below ~ 5 GeV:

$$<n_{ch}> = a + b \, \ln(W_{\gamma\gamma}/\text{GeV}). \tag{7.13}$$

In e^+e^- annihilation $a \approx 2$ and $b \approx 1.4$. For two-photon reactions one should try to determine a and b from the data.

The <u>neutral multiplicity</u> may be related to the charged multiplicity by: $\langle n_{neu}\rangle : \langle n_{ch}\rangle = 1 : 2$. No deviation from this ratio for $\pi^\circ/\pi^\pm$ has been observed yet. But note that in the analysis of one-photon annihilation data at SPEAR /163/ which also used a model with pions only, the charged particle distributions could be fitted better with the ratio $\langle n_{neu}\rangle : \langle n_{ch}\rangle = 1 : 1$. An interpretation of this result may be that this ratio accounts effectively for the production of particles with higher masses and of neutrinos.

Finally a proper <u>multiplicity distribution</u> has to be chosen. In the first analyses Poisson distributions were used. There is some arbitrariness in how to obey the constraint that the total charge has to be zero. Because of this constraint the quantity that should be Poisson distributed is in principle not n_{ch}, but rather the number of neutral pairs of charged particles, n_{pair}, or the sum $n_{pair}+n_{neu}$. The use of the formula introduced in /164/ for multiplicity distributions (KNO scaling formula) would be more motivated by the underlying physics. KNO scaling means that the multiplicity distribution can be described by a universal, energy independent function of the variable $\langle n\rangle/D$ which is the ratio of the mean value and the dispersion of the multiplicity. KNO scaling has been observed both in e^+e^- and hadronic reactions, however, with different values for $\langle n\rangle/D$. According to the VMD picture, in $\gamma\gamma$ reactions at low Q^2 $\langle n\rangle/D$ may be closer related to hadronic reactions rather than to e^+e^- scattering. In future analyses it has to be shown that KNO scaling also holds for $\gamma\gamma$ reactions and, if so, the value of $\langle n\rangle/D$ has to be measured.

It has yet to be found experimentally whether the above described model for the hadronic final state offers enough freedom to account for the structure of the data in the whole $W_{\gamma\gamma}$ range which is fitted at the same time. Another question is, whether the experiments are really sensitive to different models and what the model dependence of the results is.

7.3   Experimental Results on $\sigma(\gamma\gamma\to\text{Hadrons})$

7.3.1   A Double-Tag Analysis

Total cross section measurements in the resonance region using the double-tag method have been done by an experiment at SPEAR (at beam energies of about 3 GeV). The total cross section was measured employing the good resolution of NaI shower counters in the tagging devices (tagging angle: 55 - 150 mrad) /26/. The square of the invariant mass of the $\gamma\gamma$ system, m_x^2, was determined from the kinematics of the tagged leptons. Figure 7.5 shows as a function of m_x^2 the rate of events which had one additional charged track detected at angles between 15° and 30° from the beam. This track had to be identified as (a) a muon or (b) a pion. From these events the ratio

$$R_{\gamma\gamma} = \frac{\sigma(e^+e^-\to e^+e^-\text{hadrons})}{\sigma(e^+e^-\to e^+e^-\mu^+\mu^-)} \tag{7.14}$$

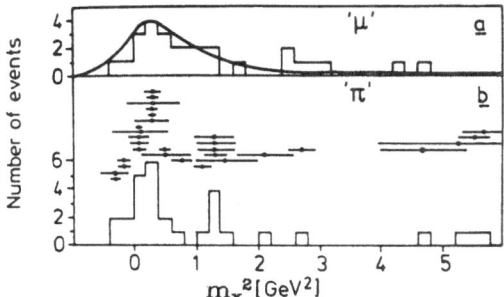

Fig.7.5.   Measurement of muon and pion production in a double-tag experiment /26/: Plotted is the event distribution as a function of the invariant $\gamma\gamma$ mass as determined from the tagging kinematics; a) events with identified muons, b) events with identified pions.

was calculated. The result for $0.07 \leq Q_1^2$, $Q_2^2 \leq 0.3$ GeV^2 and $0 \leq W_{\gamma\gamma}^2 \leq 6.6$ GeV is:

$$R_{\gamma\gamma} = 1.1 \pm 0.3 \pm 0.3.$$

To give a point of reference for this number: if the total hadronic cross section at low $W_{\gamma\gamma}$ were given by the Born approximation for $\gamma\gamma \to \pi^+\pi^-$, $R_{\gamma\gamma}$ would be at least an order of magnitude smaller. This is demonstrated in Fig.4.6 which shows the relative contributions of e, μ, and π pair production for $0.3 \leq W_{\gamma\gamma} \leq 0.7$ GeV, measured with the DM1 detector. In the SPEAR experiment the $\pi^+\pi^-/\mu^+\mu^-$ ratio would be even smaller because the detector accepted π's and μ's at smaller angles, where μ production is relatively enhanced. Since the bulk of 'π' events lies below $m_x = 0.7$ GeV, the only explanations for a large $R_{\gamma\gamma}$ value are either a large contribution from $\pi^+\pi^-\pi^\circ$ or $\pi^+\pi^-\gamma$ final states (about 10 times larger than the $\pi^+\pi^-$ Born term !) or a strong Q^2 dependence ($Q^2 \approx 0$ for the DM1 experiment, $Q^2 \gtrsim 0.07$ GeV^2 for the SPEAR experiment). It should also be noted that the inelastic Compton scattering process (Fig.2.1b) is favoured in the kinematical range accepted by the SPEAR setup. In particular, ρ production via this diagram may be an appreciable background.

7.3.2   Single-Tag Analyses

The PLUTO Analysis:
    The first measurement of the $\gamma\gamma$ total cross section for the production of hadrons was reported by the PLUTO collaboration /158/. For an integrated luminosity of 2.7 pb^{-1}, the cross section was evaluated in a $W_{\gamma\gamma}$ range up to 8.5 GeV at beam energies around 15 GeV. An electron with an energy greater than 4 GeV was required to be detected in the forward detector (23 to 55 mrad) corresponding to four-momentum transfers of the tagged photon up to $Q^2 = 0.6$ GeV^2. On the side opposite to the tag, the

electron was not allowed to be scattered under an angle exceeding 23 mrad, thus defining a quasi-real photon ('anti-tag'). In order to select multi-hadronic events, at least three particles had to be detected in the central detector, either as three charged tracks or two charged tracks and a photon shower. Because of the limited acceptance, the invariant mass of the detected hadrons W_{vis}, as reconstructed from the charged tracks and the shower energy, is on average lower than the true $W_{\gamma\gamma}$, typically by 15-20%[6] .

The detector efficiency was calculated by Monte Carlo simulation of $\sigma_{TT} + \varepsilon\sigma_{LT}$ using a multi-pion phase space model with limited transverse momenta. The p_T^2 distribution in Fig.7.6 (for all W_{vis}) was best reproduced by the Monte Carlo simulation if 75% of the events were generated with a p_T^2 dependence $\sim\exp(-5p_T^2)$ and 25% with $\sim\exp(-1p_T^2)$. The PLUTO group notes that the detection efficiency depends critically on the transverse momentum distribution. The particle multiplicity was generated according to a Poisson distribution using the average charged multiplicity $\langle n_{ch}\rangle$ = 2.0 + 1.4·ln($W_{\gamma\gamma}$/GeV) and a neutral to charged ratio of 1 : 2. These parameters were not adjusted to the data but describe the data well.

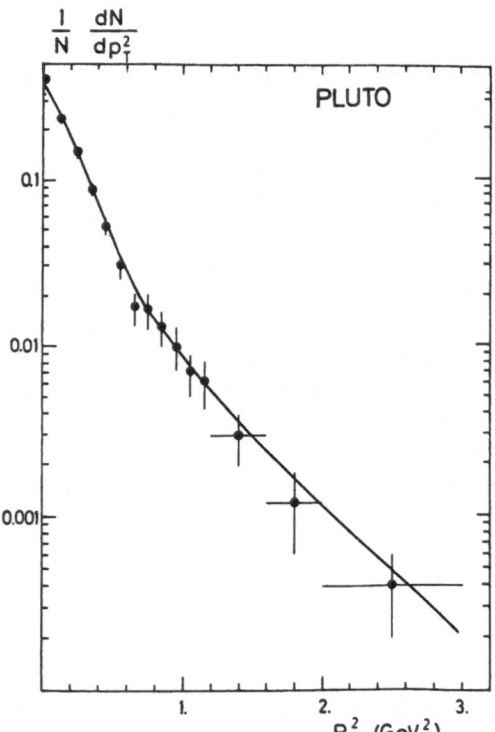

Fig.7.6.   Inclusive transverse momentum distribution of charged hadrons (PLUTO). The solid line is the prediction of the Monte Carlo simulation as described in the text.

[6]   This deviation is an average over detected events at all $W_{\gamma\gamma}$. The average is dominated by events at low $W_{\gamma\gamma}$ where events, once accepted, are more completely detected, but where the probability to detect the event at all is low. At high $W_{\gamma\gamma}$ the final state is more incompletely detected, but the event detection efficiency is higher.

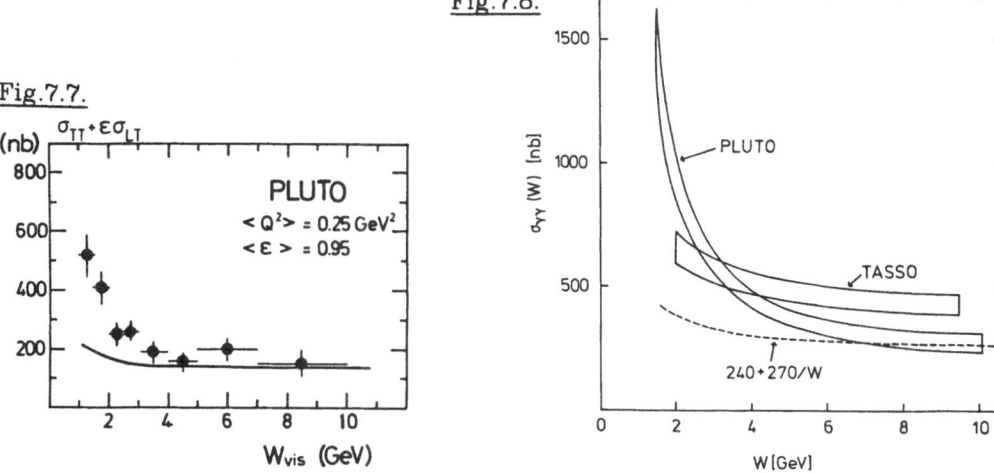

Fig.7.7.

Fig.7.8.

Fig.7.7.   The total cross section for two-photon production of hadrons as a function of the visible invariant mass at $<Q_1^2>=0.25$ GeV² (PLUTO).   The prediction of the Regge estimate is given by the solid curve.

Fig.7.8.   The extrapolated total cross section for scattering of two real photons.

The unfolding in $W_{\gamma\gamma}$ was done by fitting the W_{vis} distribution (Fig.7.7) with the following expression for the cross section:

$$\sigma = (A \cdot \sigma_{VMD} + B/W_{\gamma\gamma}^2) \cdot F_\rho(Q_1^2)$$

with $\sigma_{VMD} = (240 + 270$ GeV$/W_{\gamma\gamma})$ nb, according to (7.02) to (7.04). In Fig.7.8 the result of the fit is shown, which yielded for the parameters A and B:

A = 0.97±0.16, B = 2250±500 nb GeV².

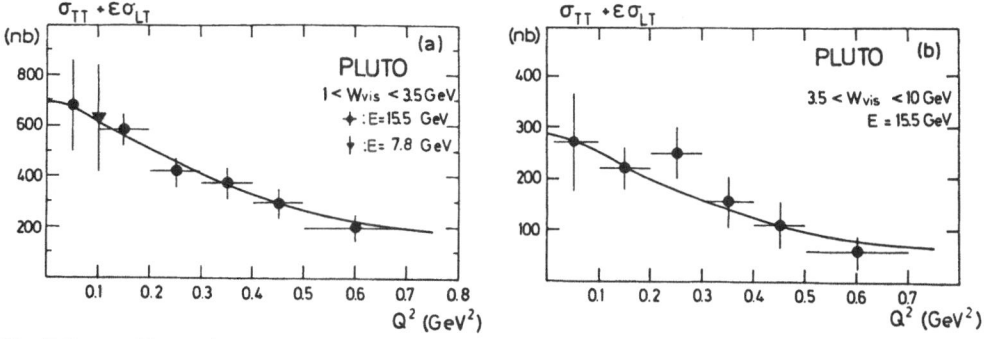

Fig.7.9.   The two-photon total cross section as a function of Q^2 (PLUTO): a) for $W_{vis}<3.5$ GeV; b) for $W_{vis}>3.5$ GeV. The solid line is the prediction from ρ meson dominance.

In Fig.7.9 the cross section is plotted as a function of Q^2 for two W_{vis} intervals and compared to the VMD prediction with a ρ pole form factor. It has been noted /165/ that the data are also consistent with the GVDM ansatz suggested in /161/. In the region $1 \leq W_{vis} \leq 3.5$ GeV, which is dominated by the $1/W_{\gamma\gamma}^2$ term, one might have expected a slower fall-off if the $1/W_{\gamma\gamma}^2$ term is connected to the box diagram.

The TASSO Analysis:

The TASSO collaboration carried out a preliminary analysis of the total cross section /159, 160/ very similar to the one described above for the PLUTO experiment. The data sample, corresponding to an integrated luminosity of 2.3 pb^{-1}, was collected under nearly the same conditions: the scattered electrons had to deposit more than 4 GeV in the forward detector at angles between 23 and 60 mrad. However, no 'anti-tagging' was required. Therefore it had to be assumed that the contribution of longitudinal photons radiated from the undetected lepton is negligible. To assure that a hadronic final state was produced, three charged tracks were required to be detected in the central detector. No neutral energy measurement was included in the analysis, which is the main difference and the main drawback compared to the PLUTO analysis.

For the determination of the $W_{\gamma\gamma}$ dependence of the cross section the measured distributions have been compared to distributions obtained from a Monte Carlo simulation of hadron production by two photons. The cross section as used in the Monte Carlo program was parametrized in the following way (including ρ pole formfactors for both photons):

$$\sigma = (X + Y/W_{\gamma\gamma})\cdot F_\rho(Q_1^2)\cdot F_\rho(Q_2^2).$$

The hadronic final state was generated in the same way as was done by PLUTO, with a p_T^2 distribution $\sim\exp(-\alpha\cdot p_T^2)$ and the average charged multiplicity $<n_{ch}> = a+b\cdot\ln(W_{\gamma\gamma}/GeV)$. In a separate step the parameter α was determined by fitting the p_T^2 distribution. The $W_{\gamma\gamma}$ dependence (X, Y) was obtained by fitting the Monte Carlo distributions of W_{vis} and n_{ch} to the experimental distributions allowing all parameters X, Y, a, b to vary independently. The best fit for the region $2 \leq W_{\gamma\gamma} \leq 9.5$ GeV yielded the following values:

$\sigma(\gamma\gamma\rightarrow$hadrons$) = (370\pm35 + (570\pm100$ GeV$)/W_{\gamma\gamma})$ nb

$<p_T>$ $\qquad = 360\pm12$ MeV, $\quad (<p_T> = \frac{1}{2}\sqrt{\pi/\alpha})$

$<n_{ch}>$ $\qquad = (2.1\pm0.12) + (1.5\pm0.07)\cdot\ln(W_{\gamma\gamma}/GeV)$

(statistical errors only) .

The differences from the PLUTO procedure are:

— All parameters other than $<p_T>$ were determined by a simultaneous fit, allowing also for correlations between the parameters.

— No $1/W_{\gamma\gamma}^2$ term was included (there was no obvious indication for it in the data).

–   The ratio of the constant to the $1/W_{\gamma\gamma}$ term was allowed to vary, whereas PLUTO fixed it at the value predicted from the Regge model.

In Fig.7.8 both the PLUTO and TASSO results are plotted. In the past, the differences of both results have often been discussed. But one has to note that, given all the systematic uncertainties involved, the agreement is not so bad for larger $W_{\gamma\gamma}$. The TASSO group found that the data do not permit a reliable cross section measurement below 2 GeV. The W_{vis} distribution below 2 GeV is strongly dominated by events having higher $W_{\gamma\gamma}$ values, while the acceptance for events with low $W_{\gamma\gamma}$ is very small.

A Study of the Problems in the Determination of the Total Cross Section:

With a larger data set, collected for an integrated luminosity of 9 pb^{-1}, the TASSO collaboration repeated the analysis and studied in great detail the influence of the different model parameters on the determination of the total cross section /168/. The data selection was similar to the previous analysis: in particular, three or more charged tracks from the hadronic final state were required to be detected and no neutrals were included. After all cuts 988 events were left in the W_{vis} range $1 \leq W_{vis} \leq 0.42 \cdot E$. The Monte Carlo model used for the unfolding of the W_{vis} distribution was structured as discussed above. The $W_{\gamma\gamma}$ dependence was parametrized with three independent terms and the Q^2 dependence was chosen according to the GVDM formula (7.11):

$$\sigma(W_{\gamma\gamma}, Q_1^2, Q_2^2) = (A + B/W_{\gamma\gamma} + C/W_{\gamma\gamma}^2) \cdot F_{GVDM}(Q_1^2) \cdot F_{GVDM}(Q_2^2). \tag{7.15}$$

The extrapolation to $Q^2=0$ with a simple ρ pole formfactor yielded a cross section about 50% higher than with the GVDM form factor. Here again the hadronic final state was generated as a multi-pion phase space with limited transverse momentum. A p_T dependence $d\sigma/dp_T^2 \sim \exp(-\alpha\sqrt{p_T^2+m_\pi^2})$ was chosen[7]. This form was found to describe the distributions at low p_T in hadronic reactions. The data seem to disfavour a Gaussian type p_T dependence $\sim\exp(-\beta\cdot p_T^2)$. The ansatz for the average charged multiplicity was taken from /169/:

$$\langle n_{ch} \rangle = n_0 + 0.0029 \cdot \exp(2.85 \cdot \sqrt{\ln(W_{\gamma\gamma}^2/0.25 GeV^2)}). \tag{7.16}$$

This formula describes e^+e^- annihilation data up to the highest available energies. In the energy range below ~10 GeV the difference from the formula used in the previous analysis is small. The ratio of neutral to charged multiplicity was fixed to 1 : 2, because the data, not including information on the neutral particles, were not sensitive to this ratio. The multiplicity distribution was chosen according to the KNO scaling formula with a ratio

---

[7]   In $\gamma\gamma$ reactions a better choice for the mass parameter in the exponential may be the ρ mass /12/.

of the mean value to the dispersion of 3.2, as was measured in e^+e^- annihilation at 9.4 GeV /170/ (see also the remarks on the choice of $<n>/D$ in Sect.7.2.3).

The unfolding procedure aimed at a determination of the parameters A, B, C, n_0 and α by fitting simultaneously the following experimental distributions with the corresponding Monte Carlo distributions:

$$dN/dW_{vis}, \quad dN/p_T{}^2, \quad dN/n_{ch}, \quad dN/dL . \tag{7.17}$$

In the last distribution $L = \sum p_z / |\sum \vec{p}|$ is the measured momentum component of the hadronic system along the beam direction, normalized to the total measured momentum. The distribution of L gives an additional check on the model assumption of production according to phase space with limited p_T. The highest sensitivity was obtained by studying the $p_T{}^2$, n_{ch} and L distributions in different W_{vis} intervals:

$$\frac{d^2N}{dW_{vis}dp_T{}^2}, \quad \frac{d^2N}{dW_{vis}dn_{ch}}, \quad \frac{d^2N}{dW_{vis}dL} . \tag{7.18}$$

The determination of the optimal set of parameters by fitting the Monte Carlo distributions to the data requires the knowledge of the Monte Carlo distributions for any set of parameter values in some region of the parameter space around the optimum. To keep the number of time consuming Monte Carlo runs as small as possible the following method was applied:

1. For each cross section term ($\sim$const., $\sim 1/W_{\gamma\gamma}$, $\sim 1/W_{\gamma\gamma}{}^2$) a set of Monte Carlo simulations was run with different settings of the nonlinear parameters (n_0, α). The distributions for any combination of the parameters A, B, C can then be obtained by linear superposition of the generated distributions.

2. In the next step each distribution was divided into bins and the bin contents were expressed as functions of the parameters by fitting a polynomial of 2nd order in the parameters to the Monte Carlo bin contents.

3. In the last step this functional form of the bin contents was used to determine with a maximum likelihood fit the optimum set of parameters by comparing the Monte Carlo simulation with the data.

This method allows one to vary the parameters continously and to determine the parameters together with their correlations.

Various fits showed that the correlations between the parameters are so large that with the given model an explicit determination of the $W_{\gamma\gamma}$ dependence of the total cross section is not possible. Monte Carlo studies proved that it was not even possible to distinguish between the constant and the $1/W_{\gamma\gamma}$ term unless one of the other parameters (multiplicity or p_T slope) was kept fixed. Since the complete $W_{\gamma\gamma}$ dependence could not be determined, an attempt was made to derive the cross section at high $W_{\gamma\gamma}$, where it could be assumed to be constant. For such a fit the data sample was restricted to events with at least four charged tracks with $W_{vis} > 3$ GeV (102 events) to minimize the acceptance biases. The following result was obtained from the fit.:

$$\sigma(\gamma\gamma \rightarrow hadrons) = 285 \pm 55 \text{ nb}$$
$$n_0 = 3.96 \pm 0.30$$
$$d\sigma/dp_T^2 \sim \exp[(6.2 \pm 0.3) \cdot \sqrt{p_T^2 + m_\pi^2}\ ].$$

A p_T^2 dependence $\sim \exp(-\beta \cdot p_T^2)$, as had been used in the PLUTO and in the previous TASSO analysis, was strongly disfavoured in this analysis.

Keeping the cross section fixed at 285 nb the 1σ errors of the parameters n_0 and α are rather tight (Fig.7.10). The 95% confidence level contour in the same plot defines the boundary for possible n_0, α combinations if the cross section is also allowed to vary. The numbers on this contour show the range of cross section values allowed by the data (for this model!) within the 95% confidence level limit. Changes of the cross section between 150 and 1300 nb indicate that the fit is not able to discriminate between different choices of parameters. Only the knowledge of either the p_T slope or the multiplicity would solve the problem. In the first TASSO

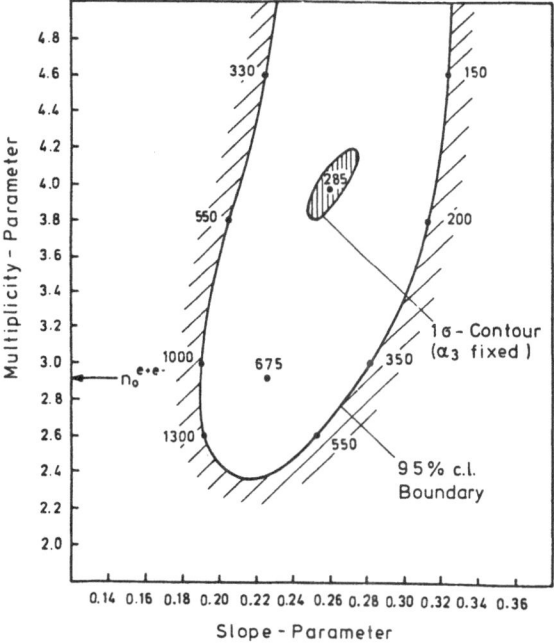

Fig.7.10. Correlation between the parameter for the transverse momentum distribution and the multiplicity parameter: The 1σ contour for fixed cross section (=α_3) and the 95% confidence level limit for the case that the cross section is allowed to vary is given together with the corresponding cross section values. See also the description in the text.

analysis, for example, the p_T slope was fixed before fitting the other parameters.

In conclusion, under the described experimental conditions the determination of the total hadronic $\gamma\gamma$ cross section seems to be impossible without assuming some a priori knowledge of the model parameters.

7.4   Discussion of the Results

The study presented in the previous section was done with a specific detector. It is clear that a higher detection efficiency for both charged and neutral particles will improve the results of the unfolding procedure. However, some conclusions are of more general validity: The study has shown that the determination of the total cross section is more complicated than assumed in the first analyses. Only a simultaneous fit of all model parameters (except they can be obtained from other sources) can reveal correlations and can lead to a significant result for the $W_{\gamma\gamma}$ dependence of the cross section. The significance of such a result is considerably increased if one can show that all distributions, especially the p_T^2 and the multiplicity distributions, are equally well described for all W_{vis} values. In particular, the determination of the p_T slope before the proper unfolding, without regarding correlations with other parameters and averaging over all W_{vis} values, was found to be dangerous.

The determination of the cross section below $W_{\gamma\gamma} \sim 2$ GeV is particularly difficult (see also the discussion in /166/). Figure 7.7 demonstrates that the large $1/W_{\gamma\gamma}^2$ term in the cross section found by the PLUTO group is required only by the data points below $W_{\gamma\gamma} \sim 2$ GeV. However, in the resonance region one has to expect special problems: the acceptance for events below 2 GeV is only ~8% /166/. On the other hand, about 85% of the events observed at $W_{vis} < 2$ GeV come from higher $W_{\gamma\gamma}$ values because of detection inefficiencies /167/. In such a situation the cross section cannot be obtained by applying simple acceptance corrections as was done in Fig.7.7, where the cross section is given as a function of W_{vis}. A proper unfolding procedure is necessary as described e.g. in /168/. Concerning the applicability of the used model to the resonance region, one has to question whether a multi-pion phase space model with limited p_T and some statistical multiplicity distribution works well enough for resonances. Another uncertainty has been mentioned in /166/: At low $W_{\gamma\gamma}$ the Q^2 extrapolation to $Q^2=0$ is especially large, about a factor of 3 in the PLUTO analysis.

Some features of the hadronic final state have been established by the described analyses. For example, we find that the p_T^2 distributions in $\gamma\gamma$ reactions at low Q^2 and low p_T have the same exponential behaviour as observed in hadronic reactions. However, at the moment, it seems to be difficult to draw final conclusions about the size and the $W_{\gamma\gamma}$ dependence of the total cross section. Given the systematic problems of such an analysis, it is desirable to have a confirmation of the PLUTO result from other experiments.

The detectors with special forward spectrometers for the detection of the $\gamma\gamma$ final state, like PLUTO 81 and PEP-9, have much better experimental conditions for total cross section measurements than the first pioneering experiments. The coverage of the forward region is important particularly because it allows a more reliable determination of the p_T dependence which is very sensitive to the acceptance corrections. Improvements can also be made by constructing more realistic models. For example, one should include different particle species and resonance formation. The model should also account for possible changes of the behaviour of the hadronic final state as a function of $W_{\gamma\gamma}$. One may consider, e.g., including in the model the measured cross sections for the production of resonances.

We expect to soon have many more results on the two-photon total cross section, especially from the PLUTO and PEP-4/PEP-9 experiments with their good forward angular coverage. Recently, the ARGUS collaboration proposed the installation of a 0° tagger at DORIS /22/. This device is supposed to achieve a resolution good enough to measure the total cross section down to the resonance region in a double-tag experiment.

8. Hard Scattering Processes in Two-Photon Production of Hadrons

8.1 General Remarks

Two-photon reactions are especially suited for testing perturbative QCD calculations because of the simplicity of the initial state of colourless, pointlike particles. However, only those kinematical regimes involving a large momentum transfer between particles participating in the reaction are accessible to perturbative calculations. Whereas in the next chapter we shall consider the case of a large momentum transfer from one photon to the hadrons (large Q^2), we want to discuss in this chapter the production of hadrons with large transverse momenta relative to the collision axis of the two photons.

The total cross section for hadron production in e^+e^- scattering via one-photon annihilation is directly related to the QED process of muon pair production: In the quark-parton model, the ratio

$$R_{1\gamma} = \frac{\sigma(e^+e^- \to \gamma^* \to \text{had.})}{\sigma(e^+e^- \to \gamma^* \to \mu^+\mu^-)} = 3 \cdot \sum_{\text{flavour}} e_q^2 \tag{8.01}$$

counts the number of contributing quark pairs weighted by the squares of their charges according to the QED diagram in Fig.8.1a. The QCD corrections to this diagram are less than 10%.

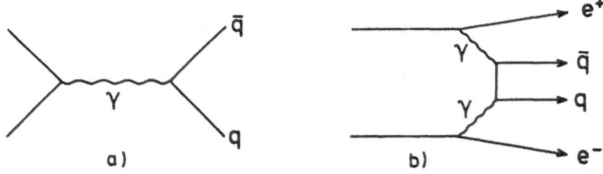

Fig.8.1. One-photon (a) and two-photon (b) production of quark pairs.

The corresponding ratio for two-photon induced hadron production

$$R_{\gamma\gamma} = \frac{\sigma(\gamma\gamma \to \text{had.})}{\sigma(\gamma\gamma \to \mu^+\mu^-)} \tag{8.02}$$

would be given by the fourth power of the quark charges if hadron production were dominated by the diagram in Fig.8.1b.

Unfortunately, the QCD corrections to this diagram seem to be large and probably not even calculable in perturbative QCD. The main difference to the one-photon case is the presence of the quark propagator. In the quark-parton model the propagator is proportional to $1/(t-m_q^2)$ where m_q is the quark mass and t is the usual Mandelstam variable for the momentum transfer carried by the exchanged quark. If $\sqrt{|t|}$ is small compared to the QCD scale parameter Λ large corrections are expected. In this regime one observes a hadron-like behaviour of the photon: the photon couples to quarks with small relative transverse momentum, having enough time to form a hadronic bound state. The vector meson dominance model provides a phenomenological description of this regime. It predicts that most of the two-photon total cross section is due to the "hadronic" part of the photon. At large $|t|$ two-photon production of hadrons is a more pointlike process because the interaction spread decreases as $1/|t|$. Above some high enough $|t|$ the diagram in Fig.8.1b should be the dominant one and $R_{\gamma\gamma}$ is expected to approach the value given by the quark-parton model with fractionally charged quarks /171/:

$$R_{\gamma\gamma} = \frac{d\sigma/dt(\gamma\gamma \to q\bar{q} \to 2 \text{ jets})}{d\sigma/dt(\gamma\gamma \to \mu^+\mu^-)} = 3 \cdot \sum_{\text{flavour}} e_q^4 = 34/27 \quad \text{(for u,d,s,c).} \tag{8.03}$$

In the Han-Nambu model of integrally charged quarks /86/ $R_{\gamma\gamma}$ has a significantly different value /85/:

$$R_{\gamma\gamma}(\text{Han-Nambu}) = 10/3. \tag{8.04}$$

Thus the determination of $R_{\gamma\gamma}$ seems to be best suited to distinguish between integrally and fractionally charged quarks. (Remember that processes involving a single photon depend only on the charges averaged over the colours which happen to be the same in both models). However, in a gauge invariant formulation of the integrally charged quark model /172/ $R_{\gamma\gamma}$ becomes dependent on the kinematics of the scattering process. In particular, $R_{\gamma\gamma} = 10/3$ is only obtained for real photons whereas the result of the fractionally charged quark model is approached when at least one of the photons has a large Q^2.

At large transverse momenta of the quarks, corresponding to large $|t|$, the process $\gamma\gamma \to q\bar{q}$ should result in events with a characteristic two jet topology, similar to one-photon annihilation, where the production of quark-antiquark pairs leads to the two jet events observed at large e^+e^- energies.

In hadronic reactions hard scattering processes have been looked for by studying the transverse momentum distribution of the produced hadrons. Whereas the dominant peripheral processes have a steeply falling p_T dependence:

$$d\sigma/dp_T^2 \sim \exp(-6p_T), \tag{8.05}$$

hard processes can be recognized by a flattening of the p_T^2 distribution, which is predicted to reach asymptotically the limit /171/:

$$d\sigma/dp_T^2 \sim p_T^{-4}. \qquad (8.06)$$

In the previous chapter we saw that in two-photon reactions the p_T distribution below ~1 GeV follows the exponential curve expected from peripheral scattering of vector mesons. At larger p_T the distribution flattens as can be seen in Fig.7.6. It was predicted /7/ that in two-photon reactions hard scattering phenomena show up in a flattening of the p_T distribution at much smaller p_T and at lower energies than in purely hadronic reactions. The observation of hard scattering processes in hadron-hadron collisions at energies similar to those of two-photon reactions at PETRA/PEP is more difficult. Because of the compositeness of hadrons, higher order processes still dominate at these energies and the detection of hard processes is complicated by the presence of the beam and target fragments.

In the following we review the experimental results on hard scattering processes in two-photon reactions and compare them to predictions obtained from the lowest order diagram for

$$\gamma\gamma \to q\bar{q}. \qquad (8.07)$$

We start with an introduction to the theoretical aspects including a discussion of higher order corrections to the Born diagram. This is followed by a presentation of the analyses of single hadron p_T distributions and of two-photon events with 2-jet topologies. Finally, a discussion of the results summarizes our present knowledge of hard scattering in two-photon reactions.

8.2   Theoretical Description of Hard Scattering Processes

8.2.1   Introduction

The authors of /171/ have shown that for pointlike constituent-constituent scattering at high energy the leading behaviour of the inclusive p_T distributions at large p_T is expected to be

$$d\sigma/dp_T^2 \sim p_T^{-4} \qquad (8.08)$$

at fixed angle Θ and fixed $x_T = 2\,p_T/\sqrt{s}$, where $\sqrt{s}$ is the center of mass energy of the reaction. Therefore, the p_T distribution is asymptotically given by:

$$d\sigma/dp_T^2 = p_T^{-4}\; f(\Theta, x_T) \qquad (8.09)$$

where f depends only on the dimensionless variables Θ and x_T. The p_T^{-4} dependence reflects the $1/s^2$ dependence of pointlike processes because for fixed angles p_T is proportional to $\sqrt{s}$. Equation (8.09) is only meaningful for reactions between composite particles involving subprocesses at different energies. In a pointlike scattering process of free particles with a fixed energy, in contrast, the variables p_T, x_T and Θ are not independent, hence the p_T distribution cannot be studied at fixed x_T and Θ. Consider for example the pointlike QED process $e^+e^- \to \mu^+\mu^-$ with the differential cross section:

$$\frac{d\sigma}{dp_T{}^2}(e^-e^- \to \mu^+\mu^-) = \frac{\alpha^2\pi}{4s^2}\frac{1 + \cos^2\Theta}{\cos\Theta} = \frac{1}{p_T{}^4}\,f(\Theta,x_T). \tag{8.10}$$

The differential cross section $d\sigma/dp_T{}^2$ for this process is plotted in Fig.8.2. Clearly, the distribution is not dominated by the factor $p_T{}^{-4}$ but rather by the function $f(\Theta,x_T)$, which peaks in fact at $p_T=\sqrt{s}/2$ (known as "Jacobian peak").

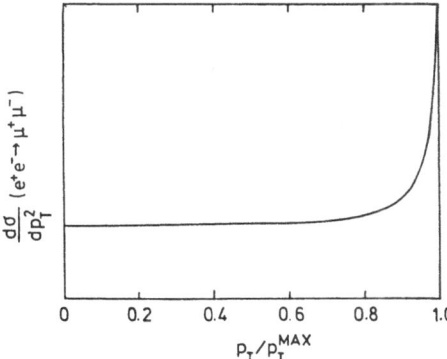

Fig.8.2.   Transverse momentum distribution of the muons in the reaction $e^+e^- \to \mu^+\mu^-$.

Equation (8.09) has interesting implications for the case of pointlike scattering of constituents in a complex reaction. In such a case the $p_T{}^2$ distribution of the final state hadrons is obtained by integrating over the elementary subprocesses. The contributing subprocesses have to be weighted by the distribution functions of the constituents (e.g. hadronic structure functions) and by the fragmentation functions (for the constituents fragmenting into the final state hadrons). In the high energy limit we write the "hard scattering expansion" /7/ for the observation of a particle C with momentum p_C in a scatter of the particles A and B:

$$E_C\,\frac{d\sigma(A+B \to C+X)}{d^3p_C} = (1/\pi)\sum_{a,b,c}\int dx_a\ dx_b\ G_{a/A}(x_a,p_T)\cdot G_{b/B}(x_b,p_T)\cdot F_{C/c}(z_c)/z_c \times$$

$$\times\,\frac{d\sigma(a+b \to c+d)}{d\hat{t}}\,\hat{s}\cdot\delta(\hat{s}+\hat{t}+\hat{u}). \tag{8.11}$$

The differential cross section for inclusive production of a particle C is expressed in terms of the <u>hard</u> subprocesses of constituent-constituent scattering a+b→c+d folded with the <u>soft</u> distribution functions G and fragmentation functions F (Fig.8.3). The variables in the integral are defined as: $x_a=\hat{t}/t$, $x_b=\hat{u}/u$ and $z_c=2E_C/\sqrt{\hat{s}}$, where s, t, u and $\hat{s}$, $\hat{t}$, $\hat{u}$ are the Mandelstam variables of the process A+B→C+X and the subprocess a+b→c+d,

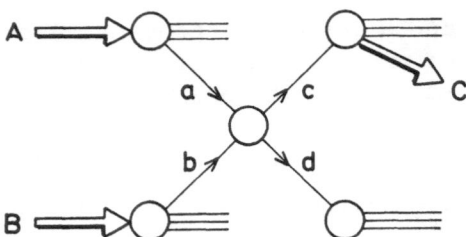

Fig.8.3. Diagram for the hard scattering expansion.

respectively. The function $f(\Theta, x_T)$ in (8.09) has to be obtained from such an integral. It depends on the distribution functions whether or not the p_T^{-4} behaviour of the hard scattering subprocesses determines the p_T dependence of the produced hadrons.

8.2.2   The Born Approximation for Hadron Production by Two Photons

In the following we discuss the production of a hadronic system X in the reaction

$$e^+e^- \to e^+e^- + X \tag{8.12}$$

via the subprocess $\gamma\gamma \to q\bar{q}$ as given by the Born diagram in Fig.8.1b. The functions G are in this case given by the probability to get photons with energy fractions ω from the electrons. In the equivalent photon approximation (2.22) we have:

$$G_{\gamma/e}(\omega) = dN_\gamma(\omega)/d\omega. \tag{8.13}$$

The differential cross section for the subprocess $\gamma\gamma \to q\bar{q}$ is:

$$\frac{d\sigma(\gamma\gamma \to q\bar{q})}{d\hat{t}} = e_q^4 \cdot \frac{2\pi\alpha^2}{\hat{s}^2} \frac{\hat{t} + \hat{u}}{\hat{t}\cdot\hat{u}} . \tag{8.14}$$

The cross section for $e^+e^- \to e^+e^- q\bar{q}$ is obtained by convoluting (8.14) with the photon fluxes (2.28) and by adding up all $q\bar{q}$ pairs (flavours and colours). Integrating in addition over the longitudinal momenta of the quarks we get /173/:

$$\frac{d\sigma(e^+e^- \to e^+e^- q\bar{q})}{dp_T^2} = R_{\gamma\gamma} \, \eta^2 \, \frac{\alpha^2\pi}{p_T^4} \, D(x_T) \tag{8.15}$$

with $\eta = (\alpha/2\pi)\cdot\log(s/4m_e^2)$.
$R_{\gamma\gamma}$ has been defined in (8.03) and p_T is here the transverse momentum of the produced quarks. For small x_T the function $D(x_T)$ is approximated by (the precise formula is given e.g. in /174/):

$$D(x_T) = (64/3)(\log(2/x_T) - 11/6). \tag{8.16}$$

At large energies the quarks turn into hadrons by forming collimated particle jets and the differential cross section in (8.15) can be interpreted as the cross section for the production of two jets. Formulae (8.15) and (8.16) show that in lowest order of the reaction $e^+e^- \to e^+e^- + 2$ jets the jet transverse momentum distribution will follow approximately a p_T^{-4} behaviour for not too large x_T. The same behaviour is expected for the QED analogue $e^+e^- \to e^+e^- \mu^+\mu^-$. The Mark J collaboration has analysed this process /42/ and found $d\sigma/dp_T^2 \sim p_T^{-4.9\pm0.2}$, in good agreement with the exact (lowest order) QED calculation of /38/ (see Fig.4.7 in Chap.4).

The single particle p_T distribution is obtained by folding (8.15) with a quark fragmentation function. That has been done in /174/ using a fragmentation function

$$x \cdot F_{h/q}(x) = 1 - a + 3a \cdot (1 - x)^2; \quad a \lesssim 1 \tag{8.17}$$

The resulting p_T dependence for inclusive hadron production via the process $e^+e^- \to e^+e^-q\bar{q} \to e^+e^- +$ hadrons is:

$$\frac{d\sigma}{dp_T^2} = R_{\gamma\gamma} \; \eta^2 \; \frac{\alpha^2\pi}{p_T^4} \; \frac{64}{6} \; (1 - a/2) \cdot (\log(2/x_T) - 7/3 + ...). \tag{8.18}$$

Comparing this expression to (8.15) we see that the p_T behaviour of the produced hadrons at medium x_T is similar to that on the quark level.

In order to compare the experimental data to the prediction in (8.15) or (8.18) one has to isolate the Born process by selecting events with two jets at large angles. Whether this procedure is meaningful depends on the answers to the following questions: 1) are there other processes which have the same event topologies and 2) are the QCD corrections reasonably small ? These questions will be discussed in the following sections.

8.2.3 Other Hard Scattering Processes

The process $\gamma\gamma \to gg$ (g=gluon) gives a contribution to the 2-jet cross section via the quark loop in diagram a) of Fig.8.4. This diagram has been estimated to contribute about 10% of the $\gamma\gamma \to q\bar{q}$ rate /174/. The gluons are linearly polarized and a measurement of polarization effects may be possible at LEP energies (the planned e^+e^- storage LEP is designed to have in the final stage beam energies up to 100 GeV). The gluon-jet contribution can be enhanced with longitudinally polarized electron beams /174/.

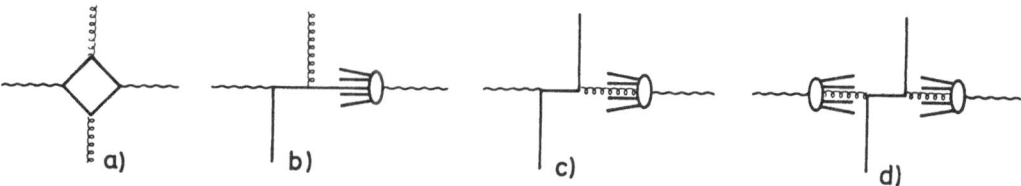

Fig.8.4. Diagrams for jet production by two photons.

Processes with 3 and 4 jets in the final state have to be considered as well although they are in principle distinguishable from the 2-jet topologies (Fig.8.4b,c,d). Since only 2 of the jets will be found at large p_T they may look very much like 2-jet events in the detector. The processes with 3 jets can be viewed as hard scattering of a bare photon off the parton content of the other photon. The parton content of a photon is given by its structure function. In diagram b) of Fig.8.4 the hard scattering process is $\gamma q \rightarrow qg$. A similar process is shown in Fig.8.4 c): $\gamma g \rightarrow q\bar{q}$. Both 3 jet processes have two large p_T jets and one small p_T jet, the 'beam pipe jet' /175/. Experimentally beam pipe jets are difficult to detect. The specialized detectors, like PLUTO 81 and PEP 9, which cover the forward direction down to 5°, may be able to provide the necessary experimental information on this event type.

The next more complicated configuration contains four jets (an example is given in Fig.8.4d). Several processes contribute, like $q\bar{q} \rightarrow q\bar{q}$, $q\bar{q} \rightarrow gg$, $gg \rightarrow q\bar{q}$, etc.. An interesting feature of the two-photon jet cross section has been noted in /177/: According to (8.11) the 3 and 4 jet cross sections involve convolutions with the photon structure function which has in leading order the behaviour $\sim 1/\alpha_s(p_T^2)$. The subprocesses, like $\gamma q \rightarrow qg$, have an $\alpha_s(p_T^2)$ factor which compensates the leading α_s dependence of the structure function. Thus the 3 and 4 jet cross sections have the same leading order behaviour as the Born cross section.

In Fig.8.5 the relative contributions of the jet cross sections are plotted as a function of p_T^{jet}. The experimental analyses have been carried out for p_T^{jet} values up to 6 GeV. According to Fig.8.5, in this region the other contributions, which have a different p_T behaviour than the Born term, are still important.

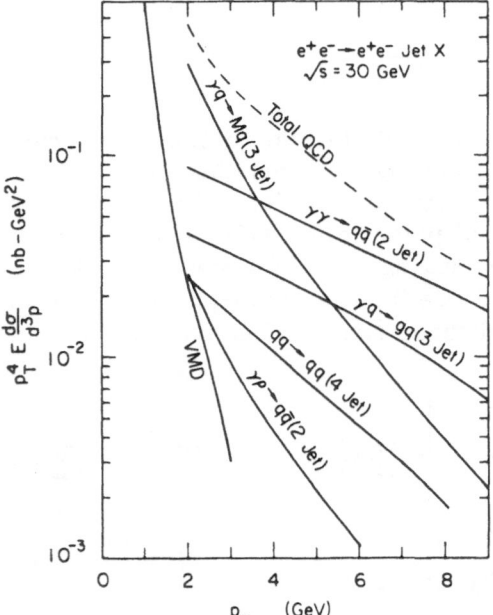

Fig.8.5. Contributions to the cross section for $e^+e^- \rightarrow e^+e^- + jet + X$ (from /7/).

8.2.4   Higher Twist Contributions

At presently available energies and transverse momenta higher twist processes, which are characterized by a faster fall-off at large p_T than the lowest order processes, could still be dominant. In /7/ it has been suggested that at current energies the leading mechanism at large p_T would be the CIM (constituent interchange model) process (Fig.8.6):

$$\gamma q \to \pi q. \tag{8.19}$$

In terms of the "hard scattering expansion" (8.11) the hard scattering subprocess includes in this case a non-elementary field, namely the pion. From dimensional counting, i.e. counting the elementary fields involved in the subprocess, a p_T^{-8} dependence is expected for the pion.

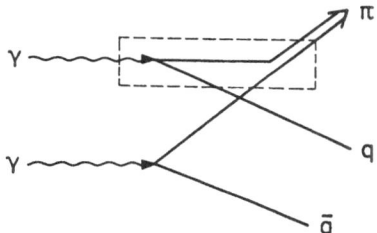

Fig.8.6.   Diagram for the CIM process $\gamma q \to \pi q$.

The importance of the CIM contribution has been questioned. The authors of /176/ find that it is suppressed by two orders of magnitude compared to the lowest order processes. However, this does not exclude the possibility that other higher twist contributions are important. Whether they are important or not has to be answered experimentally.

8.2.5   QCD Corrections to the Two-Jet Cross Section

The comparison of $R_{\gamma\gamma}$ as determined from the two-jet cross section to the lowest order quark model prediction is only meaningful if the higher order corrections are small. For one-photon annihilation the QCD corrections to $R_{1\gamma}$ as defined in (8.01) have been calculated up to second order and are found to be small. In the case of $R_{\gamma\gamma}$ we can only hope for small corrections if we restrict ourself to large transverse momenta of the hadrons or jets (see discussion above). In /178/ the virtual gluon and gluon bremsstrahlung corrections to the two-jet cross section were calculated according to Fig.8.7. The virtual together with the soft and collinear corrections were calculated in terms of Sterman-Weinberg parameters δ and ε, which define 'soft' and 'collinear' bremsstrahlung, ε being the energy fraction outside a cone of opening angle 2δ /179/.

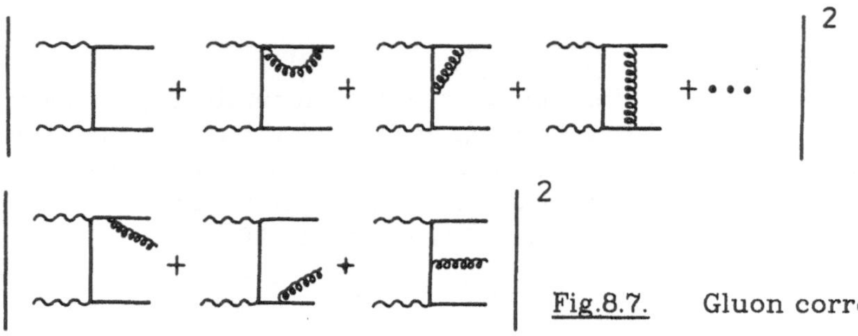

$$\left|\;\square + \square + \square + \square + \cdots\;\right|^2$$

$$\left|\;\square + \square + \square\;\right|^2$$

Fig.8.7.   Gluon corrections to $\gamma\gamma \to q\bar{q}$.

The (ε,δ) independent terms give a correction of −11% to the parton model prediction for α_s=0.3, $W_{\gamma\gamma}$=30 GeV and p_T^{jet}>4 GeV. This is a reasonably small correction. A problem arises from the (ε,δ) dependent terms. From an experimental point of view, cross section formulae which depend on (ε,δ) are inconvenient because the total $\gamma\gamma$ energy is usually not measured and the energy fraction within the δ cone is unknown. Since the total cross section must be independent of ε and δ, the (ε,δ) dependent terms have to be cancelled by corresponding terms from hard acollinear bremsstrahlung. However, the evaluation of the bremsstrahlung cross section $\gamma\gamma \to q\bar{q}g$ leads to a conceptual problem: In the configuration of Fig.8.8 one of the photons splits into a collinear quark pair; scattering with the second photon then leads to a quark and a gluon at large angles. For vanishing quark masses this diagram has a singularity $\sim\log(p_T^2/m_q^2)$, i.e. the corrections due to gluon bremsstrahlung depend on the mass scale for the quarks.

The splitting of the photon into quarks is contained in the structure function of the photon. The leading order QCD evaluation of the photon structure function for light quarks /8/ suggests that a cut-off mass M_c of the order of the QCD scale parameter Λ has to replace the quark mass (for light quarks). The authors of /178/ calculated the corrections to the Born cross section for inclusive jet production, i.e. for $e^+e^- \to e^+e^-$+jet+X. In Table 8.1, taken from /178/, the relative change of the Born cross section σ^0 for p_T^{jet} larger than a minimum p_T is given as a function of M_c. The list contains calculations for two different beam energies which are available at PETRA and PEP (15 GeV) or will be available at the planned storage ring LEP (90 GeV). The corrections are large, except for the region of large p_T where the rate is low. E.g., in the single-tag data of TASSO /180/ only 3

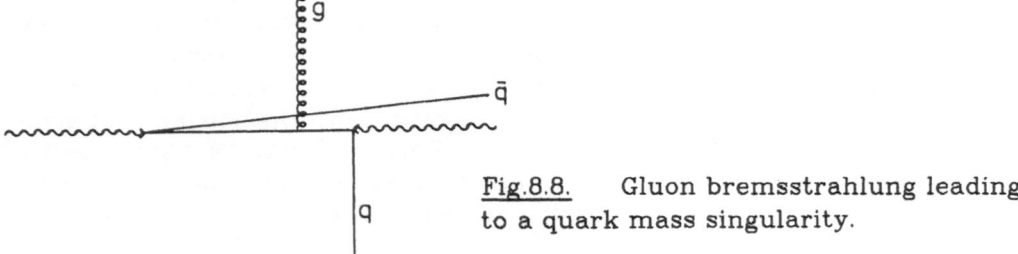

Fig.8.8.   Gluon bremsstrahlung leading to a quark mass singularity.

Table 8.1.   Corrections to the Born cross section for $e^+e^- \to e^+e^- +$jet$+$X.: σ° is the Born cross section for $e^+e^- \to e^+e^- +$jet$+$X ($p_T^{jet} > p_T^{min}$) for E$=$15 GeV and 90 GeV. σ' are the $O(\alpha_s)$ corrections for the cut off parameter M_c (taken from /178/).

p_T^{min} [GeV] →	E = 15 GeV			E = 90 GeV				M_c [GeV]
	4	6	8	12	18	24	36	
$(\sigma^\circ + \sigma')/\sigma^\circ$	1.39	1.21	1.07	1.56	1.48	1.38	1.30	0.3
	1.23	1.13	1.03	1.53	1.42	1.38	1.26	0.5
	1.11	1.03	-	1.46	1.39	1.33	1.21	0.8
σ°(pb)	15.4	2.8	0.6	6.5	4.5	1.6	0.3	

events are above $p_T^{jet}=3$ GeV. At PETRA/PEP energies the corrections are strongly dependent on M_c and thus are unreliable, whereas at LEP energies the corrections are rather independent of M_c but large.

It has been pointed out in /181/ that the corrections to the Born approximation for the inclusive jet cross section as presented above contain parts of the three jet cross section. If one were able to separate 2- and 3-jet topologies, the corrections could become much smaller. The author of /181/ shows that using (ε, δ) cuts with $\varepsilon, \delta \ll 1$ for the definition of 2-jet events, the $O(\alpha_s)$ corrections can be kept below about 20%. Calculating the corrections to the 2-jet cross section he adds those 3-jet contributions which fulfill the (ε, δ) cuts (4-jet configurations are negligible). According to well known theorems the sum of all diagrams leading to the same degenerate state has to be free of singularities. Therefore in these calculations, the $O(\varepsilon, \delta)$ terms which contain the mass singularities of Fig.8.8 are dropped. This is justified with the argument that the splitting of the photon into a collinear quark pair according to Fig.8.8 is included in the structure function and thus included in the 3-jet cross section. Two examples are given in /181/ for the $O(\alpha_s)$ correction δ_T obtained for the production of two jets at 90° (the definition of δ_T is: $\sigma = \sigma^\circ(1+\delta_T)$):

$$\delta_T = -0.16 \quad \text{for} \quad \varepsilon = 0.15, \ \delta = 10°$$
$$\delta_T = \ \ 0.19 \quad \text{for} \quad \varepsilon = 0.20, \ \delta = 10°.$$

These corrections appear to be fairly small. However, despite the apparently small absolute corrections, the change of δ_T by 0.35 with the relatively small change of ε from 0.15 to 0.20 seems to indicate instabilities. It has also been noted /182/ that in this calculation $O(\varepsilon, \delta)$ pieces have been neglected which could give finite, but a priori unknown, contributions to the corrections. In addition, the corrections given in this form may not be easily applicable to experiments. As mentioned before, the fraction ε of the usually unknown $\gamma\gamma$ energy is difficult to determine. Furthermore, the angles δ, which are required to keep the corrections controllable, are too small to be realistic in view of fragmentation effects. The experimenter would rather like to have the jet cross sections given in terms of $(\tilde{\varepsilon},\delta)$ cuts, where $\tilde{\varepsilon}$ is a fraction of the e^+e^- energy (this requires a convolution with the photon spectra) and where the jet opening angle δ is large enough to comfortably contain the jet.

To avoid the problems introduced by the (ε,δ) cuts the authors of /183/ calculate the $O(\alpha_s)$ corrections to the inclusive single hadron cross section at polar angles around 90°. The principles of the procedure followed in these calculations are briefly listed below:

1.   calculate the full $O(\alpha_s)$ corrections to the Born process;
2.   evaluate the multi-jet contributions to $O(\alpha_s)$ with the full singularity structure and subtract them from the Born result;
·3.   the comparison of the resulting cross section, which is now free of singularities, to the 0th order Born cross section gives the $O(\alpha_s)$ corrections.

With this procedure the perturbative corrections to the Born approximation are found to be small and furthermore the Born process seems to give a good description of the complete cross section, i.e. contributions from multi-jet processes are found to be small.

Two remarks should be made to these calculations: The corrections were obtained for a fixed $\gamma\gamma$ energy, whereas the measurements are done at fixed e^+e^- energy and the $\gamma\gamma$ energy is usually not determined. Secondly, the corrections were derived in the asymptotic limit and no minimum energy or minimum transverse momentum is specified where these corrections may become applicable to the data. We know from experiment that such a specification is necessary because the low p_T ($p_T < 1$ GeV) regime is not at all dominated by the Born process.

The results of the three different approaches to calculate the $O(\alpha_s)$ corrections to the Born process $\gamma\gamma\rightarrow q\bar{q}\rightarrow$hadrons can be summarized as follows:

—   The corrections to the inclusive jet cross section (i.e. not requiring the identification of a 2-jet configuration) are large and therefore probably unreliable.
—   The corrections to the 2-jet cross section, where the 2 jets are defined by (ε,δ) cuts, can be made small, by choosing (maybe for practical purposes unrealistically) small ε, δ values. However, in this calculation finite $O(\varepsilon,\delta)$ pieces have been dropped which may not be negligible.
—   The perturbative corrections to single hadron production at large p_T via the Born process are found to be very small and the complete cross section is found to be well represented by the Born process. But this work does not indicate the energy and p_T range in which the corrections are reliably computable.

We hope that the applicability of the QCD calculations will become more clear as time goes on. A lot of experimental information has still to be digested by the theorists. For example, the Q^2 dependence of the QCD corrections has not yet been discussed. Experimentally, one finds that with increasing Q^2 the Born process gives an increasingly better description of the data. A similar conclusion for quasi-real photons in the limit of large p_T of the hadrons is not yet possible. Here we may have to wait for LEP energies, for which in addition the analysis of jet topologies should become easier.

8.3 Experimental Results

Experimentally, hard scattering processes in two-photon reactions are investigated by studying the large-p_T behaviour of single hadron inclusive cross sections and by searching for events with jet topologies. The main goals of these experiments are, first of all, to establish the existence of hard processes and then to make quantitative comparisons to the theoretical predictions. Particularly interesting is a test of the relation (8.03) which requires a measurement of the Born process

$$\gamma\gamma \to q\bar{q} \to 2 \text{ jets.} \tag{8.20}$$

The first analyses of this type have been reported by the JADE and TASSO collaborations. In both cases the single-tag method was applied leading to $<Q^2> \approx 0.3$ GeV^2 for the tagged photon. Recently preliminary results were reported /81, 184/ which extend the Q^2 range considerably: the PLUTO group presented single-tag data for a Q^2 range from about 0.1 to 20 GeV^2 and the TASSO and CELLO groups analysed no-tag data (quasi-real photons).

8.3.1 The Experimental Method

The measurement of hard processes, especially the detection of jets, requires the selection of events with large $\gamma\gamma$ invariant masses. The suppression of the one-photon annihilation background is a major problem in these experiments. Without tagging, a substantial amount of one-photon background has to be subtracted by using Monte Carlo simulation. Figure 8.9 shows the distribution of the sum of the momenta (normalized to the beam momentum p_b) of the charged particles of an event as measured with the TASSO detector. The plotted events have at least one particle with a transverse momentum exceeding 1.5 GeV (this requirement suppresses the low energy two-photon events). For the study of hard processes TASSO used the region $\sum |\vec{p}| < 0.4 \cdot p_b$.

In the past experimenters preferred to use the single-tag method to reduce the large one-photon background. But even with tagging, some background from one-photon annihilation events remains. Radiated hard photons can fake a tag in the forward detector after converting in the beam pipe.

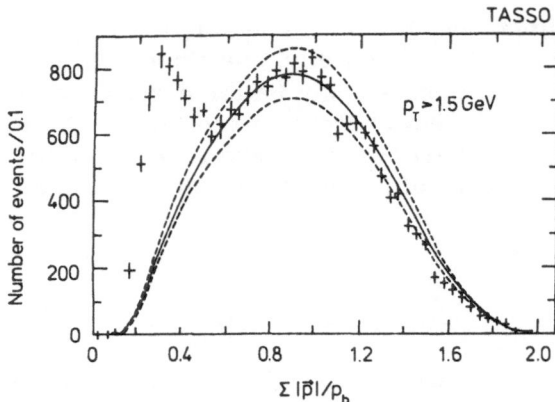

<u>Fig.8.9.</u>   Distribution of the sum of the momenta of the measured charged tracks of an event normalized to the beam momentum (TASSO).   The events are required to have at least   one track with $p_T > 1.5$ GeV.   The solid line is the expected   contribution from one-photon annihilation events   (calculated with an uncertainty indicated by the dashed curves).

For the analysis of jet topologies the experiments exploited the experience gained in recent years from the analyses of jets produced in one-photon annihilation reactions. Two-jet events from $\gamma\gamma$ reactions have in general a much lower center of mass energy than those from one-photon annihilation. Since in two-photon reactions the laboratory and center of mass systems do not coincide the jets will in general be acollinear, with an angle between the jet axes resulting from the boost of the $\gamma\gamma$ system. If the event is almost completely detected, one can boost the hadrons into their center of mass system and then apply the usual algorithms to search for two collinear jets (e.g. using sphericity, thrust etc.). Otherwise, the jet searching algorithm has to account for the acollinearity. One method is the generalized thrust algorithm: Jets are defined as those two groups C_1 and C_2 of particles for which the expression

$$T_2 \; = \; \left|\sum_{C_1}\vec{p}\right| \; + \; \left|\sum_{C_2}\vec{p}\right| \tag{8.21}$$

has a maximum. Jets can also be defined by a cluster algorithm, which does not force the events to have exactly two jets /185/. Monte Carlo studies have shown that both methods are effective in finding jets from the Born process (8.20) and reproduce the jet axes with reasonable resolutions.

8.3.2   The TASSO Analysis

For the analysis of hard scattering processes /180/ the TASSO group used the same single-tag data which were used for the total cross section analysis discussed in Chap.7. The data correspond to an integrated luminosity of 9 pb^{-1}. The average Q^2 of the tagged photons was $<Q^2> \approx 0.35$ GeV^2. In Fig.8.10 we show the $d\sigma/dp_T^2$ distribution of charged tracks plotted versus p_T. For $p_T > 500$ MeV the distribution is not affected by detector inefficiencies. The average $W_{\gamma\gamma}$ for these data has been estimated by Monte Carlo calculations to be about 6.5 GeV. Background processes, such as two-photon QED and one-photon annihilation reactions, have been subtracted statistically. The one-photon annihilation background above $p_T=1$ GeV is about 20%.

At low p_T values the distribution shows the expected exponential behaviour; above 1 GeV the distribution flattens off and approaches a p_T^{-4} behaviour. In fact, a fit of the form

$$A \cdot \exp(-B \cdot p_T) + C \cdot p_T^{-n} \qquad (8.22)$$

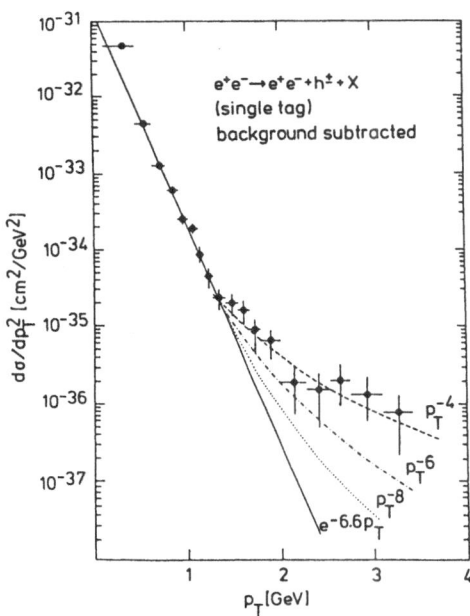

Fig.8.10.   Transverse momentum distribution for inclusive hadron production by two photons (TASSO).

yields the result $n=3.9\pm0.6$ and $B=-7.4\pm0.3$ GeV^{-1}. This result depends crucially on the requirement that there is a smooth transition between the small and large p_T region. In particular, it is not possible with the given statistics to establish the approximate p_T^{-4} behaviour by fitting only the large p_T data. Considering these systematic uncertainties, the data are compatible with both the CIM prediction of p_T^{-6} (see Sect.8.2.4) and the quark-parton prediction of p_T^{-4}. In hadronic reactions the p_T slopes observed above $p_T \sim 1$ GeV are much steeper. In the reaction $pp \rightarrow \pi^{\pm}+X$ data from the pp storage ring ISR show the following p_T dependence /190/:

$$E \, (d\sigma/dp_T^2) \sim (1-x_T)^9 \cdot p_T^{-8}. \qquad (8.23)$$

This form yields, at center of mass energies comparable to the two–photon data, an effective p_T dependence $\sim p_T^{-12}$. Thus, the production of hadrons with large p_T is relatively more abundant in two–photon reactions than in hadronic reactions. We interpret this observation as evidence for a pointlike coupling of the photon to the constituents of hadrons.

For the quantitative comparison with the Born prediction a search for 2-jet events was carried out. The jets were defined using the generalized thrust algorithm as defined above. The following requirements were specific to the jet analysis /180/:

- $\geq$4 charged tracks per event,
- 3 GeV $\leq W_{vis} \leq 0.35 \cdot \sqrt{s}$,
- $\geq$2 particles per jet,
- at least 1 jet with $p_T^{jet} \geq 2$ GeV.

After these cuts 43 events were left with an average $W_{\gamma\gamma}$ of $\sim$9.5 GeV, as determined from a Monte Carlo simulation. The plots in Fig.8.11 indicate that the selected events have jet topologies. In a) the distribution of the transverse momenta q_T of the particles with respect to the reconstructed jet axis is plotted and compared to the expectation from jet and VMD models. In the jet models the events are generated according to the QED cross section for $\gamma\gamma \to q\bar{q}$. The estimated uncertainty resulting from the quark fragmentation is indicated by the hatched band. In the VMD model the events are generated with limited transverse momenta. "Extended VMD" means that two different exponential slopes for the p_T distribution are used to account roughly for the observed p_T distribution (Fig.8.10). In b) the average sphericity for both jets (each sphericity computed with respect to the correponding jet axis) is compared to the predictions of the two models. The plots show that the data are consistent with a jet inter-

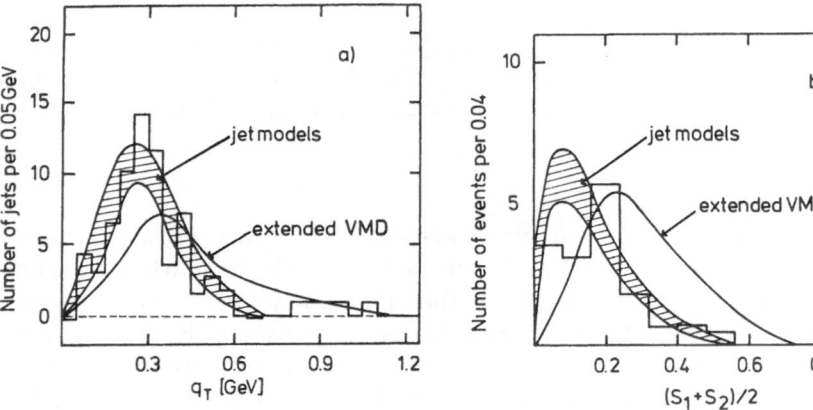

Fig.8.11.   Demonstration of jettiness of candidate events for $\gamma\gamma \to q\bar{q}$ (TASSO) : The measured distributions of a) the average transverse particle momentum with respect to the jet axis for each jet and b) the average sphericity of both jets are compared to a jet model and an "extended VMD" model (see text).

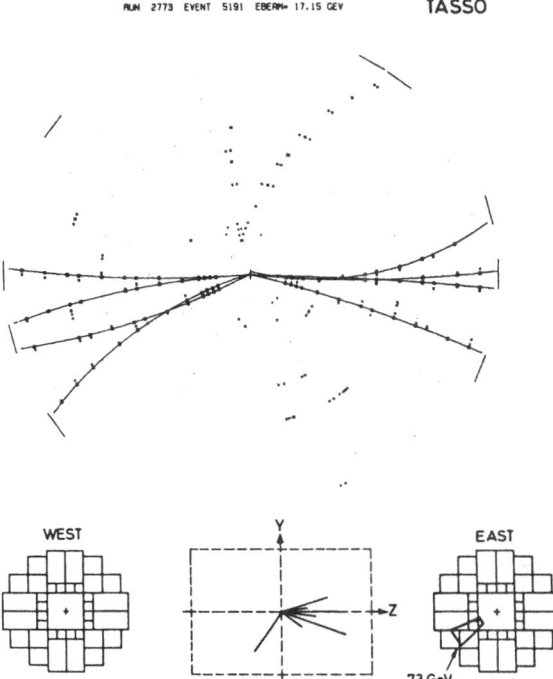

WEST   Y   EAST

                          73 GeV

Fig.8.12.   Singly tagged two-photon event with 2-jet topology: The
top part shows a view of the event in the plane perpendicular to the
beam, while the bottom part shows the event in a plane containing the
beam. The tag, indicated in the east forward detector, points in the
same direction as the event boost (east = positive z axis).

pretation. A nice example of a 2 jet event produced by two photons is
shown in Fig.8.12. This event is boosted in the same direction in which the
detected electron is scattered. Such a kinematical situation is unlikely for
a radiative one-photon event faking a tag via conversion of the radiated
photon.

   The distribution of the squared jet transverse momenta, as determined
from the detected particles, is plotted in Fig.8.13 and compared to a VMD
model and to the absolute prediction for $\gamma\gamma \to q\bar{q}$ with $R_{\gamma\gamma} = 34/27$. The
hatched band reflects mainly the uncertainties due to the quark fragmen-
tation.

   At large p_T the data lie far above the VMD curve. The prediction from
the Born diagram   approaches the data points with increasing p_T from below. For $p_T^2(\text{jet}) > 5$ GeV^2 the data are still a factor of 2 higher than the
model prediction.

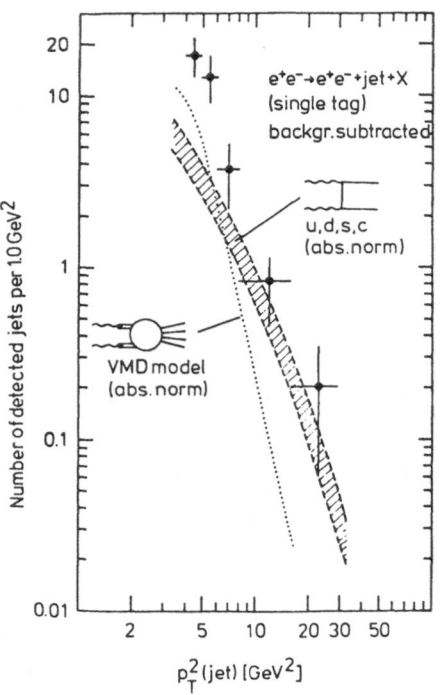

Figure content labels: Number of detected jets per 1.0 GeV², p_T^2 (jet) [GeV²], $e^+e^- \rightarrow e^+e^- + jet + X$ (single tag) backgr. subtracted, u,d,s,c (abs. norm), VMD model (abs. norm)

Fig.8.13. Transverse momentum distribution of jets in the reaction $e^+e^- \rightarrow e^+e^- + jet + X$ (TASSO): The data are compared to Monte Carlo calculations using the Born approximation for $e^+e^- \rightarrow e^+e^- q\bar{q}$ with different fragmentations (shaded band) and the vector meson dominance model (dotted line).

8.3.3   The JADE Analysis

The JADE collaboration has also used the single-tag method in their jet analysis /189/. An electron with an energy of more than 4 GeV has been required in the forward detector (34 to 75 mrad). In addition, at least 2 charged particles and at least 4 particles (charged or neutral) in total had to be detected in the central detector. To remove events with small $W_{\gamma\gamma}$, which do not contribute to the jet cross section, the energy E_{vis} measured in the central detector had to exceed the momentum component of the detected system in the beam direction by 3 GeV. One-photon annihilation events have been suppressed by requiring $E_{vis}<20$ GeV. 538 events survived these cuts corresponding to a luminosity of 9.7 pb⁻¹. The remaining background comes mainly from one-photon annihilation. The following upper limits for this background have been derived assuming that for $15<E_{vis}<20$ GeV all selected events come from one-photon annihilation: 2% for $3<E_{vis}<10$ GeV and 20% for $10<E_{vis}<15$ GeV. With the cluster method 119, 104 and 9 events were found with 1-jet, 2-jet and more than 2-jet topologies, respectively.

In Fig.8.14 the inclusive particle p_T distribution[8] is plotted for

a) all selected events (538)
b) all 2-jet events (104)
c) all 2-jet events having at least one jet with $p_T^{jet}>2$ GeV (42).

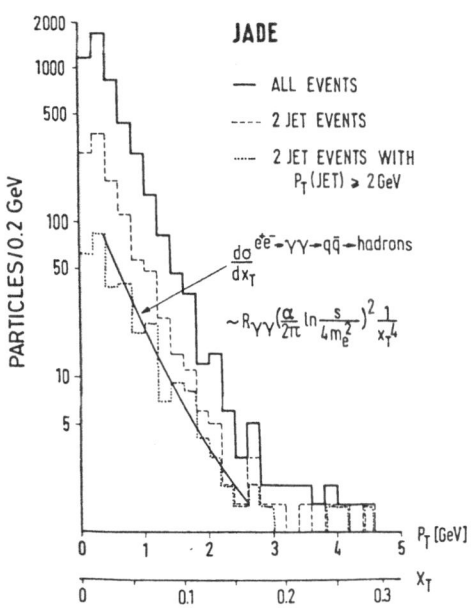

Fig.8.14. Transverse momentum distribution for inclusive hadron production by two photons compared to the expectation for $e^+e^- \to e^+e^-q\bar{q} \to e^+e^-+2$ jets (JADE).

The shape of the distribution of the last sample is compared to the prediction for $\gamma\gamma \to q\bar{q}$ with quark fragmentation à la Field-Feynman /187/ (the normalization is adjusted to the data). The slope of the distribution is well described by the data. That is not the case for the whole sample (a) or for the whole 2-jet sample (b), which have a steeper distribution than predicted by the quark model. However, the JADE group remarks that in all other distributions which have been checked, such as multiplicity, invariant $\gamma\gamma$ mass and thrust, the whole event sample (a) agrees well with the quark-jet model prediction. That may just mean that events with low-p_T jets cannot be identified as such. However, a more relevant question is whether the 42 events with high-p_T jets can be clearly identified as two-jet events.

The x_T^{jet} distribution ($x_T^{jet} = p_T^{jet}/E$, E = beam energy) for these 42 events from sample c) is plotted in Fig.8.15. From the Born cross section for $\gamma\gamma \to q\bar{q}$ with $R_{\gamma\gamma} = 34/27$ only 20 events are expected. With increasing x_T^{jet} the disagreement with the Born approximation seems to become smaller.

---

8   Note that in the JADE analysis the transverse momentum p_T is defined in a somewhat unusual way, namely with respect to the direction of the movement of the hadronic system in the laboratory. Usually, p_T is defined with respect to the $\gamma\gamma$ axis in the $\gamma\gamma$ center of mass system, which can be approximated for small Q^2 by the beam direction. In most cases both definitions nearly coincide because the $\gamma\gamma$ system tends to have a boost in the direction of the beam axis. It has been checked by the JADE group that the standard definition of p_T leads to the same results as the ones presented in the published paper /186/.

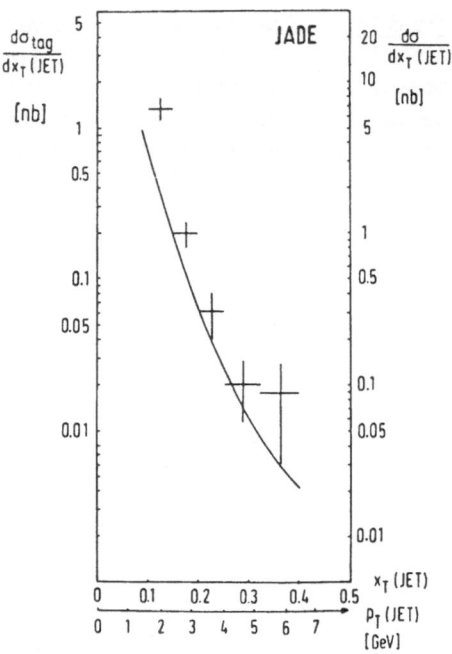

Fig.8.15. Transverse momentum distribution of jets in the reaction $e^+e^- \to e^+e^- + jet + X$ compared to the absolute prediction for fractionally charged quarks (JADE).

8.3.4   The Q^2 Dependence of the Transverse Momentum Distributions

A very interesting topic is the Q^2 dependence of the p_T distributions for single hadrons and jets. One might expect that with increasing Q^2 the pointlike processes, characterized by a p_T^{-4} behaviour, would become more prominant compared to higher order processes. Concerning the determination of $R_{\gamma\gamma}$, one has to note that the gauge invariant version of the integrally charged quark model predicts $R_{\gamma\gamma}$ to change with Q^2 such that the difference to the prediction for the fractionally charged quark model is large at small Q^2 and small at large Q^2.

The first analyses of large-p_T processes have been carried out under the single-tag condition ($<Q^2> \approx 0.3$ GeV^2 for the tagged photon) in order to ease the separation of the hard $\gamma\gamma$ processes from one-photon annihilation events. In recent analyses of the PLUTO /81, 184/, TASSO /188/ and CELLO /184/ groups the Q^2 range has been extended to both very small Q^2 values (no-tag) and very large Q^2 values. The PLUTO and CELLO analyses are still preliminary.

The PLUTO group has studied single-tag data taken with the small angle tagger ($0.1 < Q^2 < 1$ GeV^2) and with the large angle tagger ($1 < Q^2 < 20$ GeV^2). In Fig.8.16 the single hadron inclusive p_T^2 distributions are plotted for the two cases. For both plots $W_{vis} > 4$ GeV has been required, where W_{vis} has been reconstructed from the measured neutral and charged particles. The transverse momentum p_T has been measured with respect to the reconstructed $\gamma\gamma$ axis in the $\gamma\gamma$ center of mass system. The data are compared to a VMD model and to the Born approximation for $\gamma\gamma \to q\bar{q}$ including u, d, s,

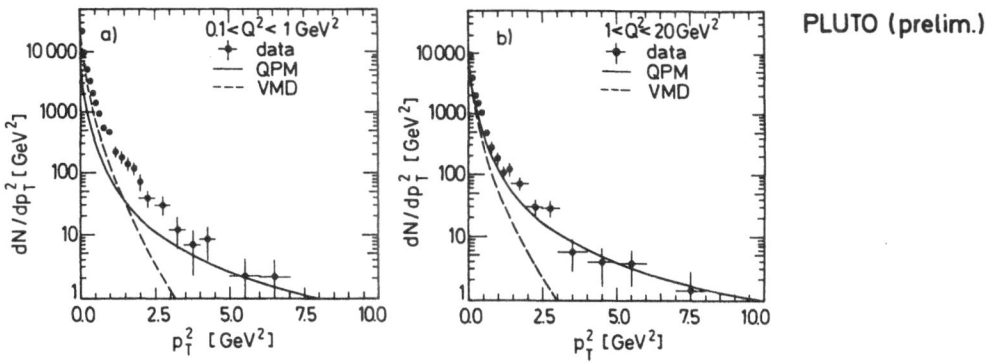

<u>Fig.8.16.</u>   Transverse momentum distribution of single charged hadrons for two Q^2 ranges (a, b) of the tagged photon (PLUTO) : The measured data are compared to predictions from a VMD model with exponential p_T dependence (dashed curve) and to the prediction from the quark Born diagram (QPM) with Field-Feynman fragmentation (solid curve).

c quarks. The small Q^2 data lie far above the Born prediction at low $p_T{}^2$ possibly approaching this curve at large $p_T{}^2$, whereas for the high Q^2 data the excess over the Born term is everywhere much smaller and the data are consistent with the Born term above $p_T{}^2 \approx 2$ GeV². The difference between the small Q^2 data and the Born prediction seems to increase as one goes to even smaller Q^2. In an analysis of no-tag data (Q^2 dominantly very small) /188/ the TASSO group has shown that for $1.5 < p_T < 3.0$ GeV the measured inclusive $p_T{}^2$ distribution exceeds the expectation from the Born term by about a factor of four (see Fig.8.17).

The question arises whether the event topologies are different in different regions of Q^2 and p_T. One might expect that in regions where the Born term dominates the 2-jet topology is more pronounced than in regions where higher order processes are more important. The PLUTO group therefore carried out a jet analysis of the small and large Q^2 data. They used the generalized thrust method as described above analysing each event as a 2-jet event. Somewhat surprisingly, one finds that in any region of Q^2 and W_{vis} the average event looks "jetty", independent of Q^2 and W_{vis}. However, a plot of $\cos\Theta^{Jet}$, where Θ^{Jet} is the angle between the thrust axis and the $\gamma\gamma$ axis in the center of mass system of the hadrons, reveals that for low Q^2 the jet production is more strongly peaked in the forward direction than predicted from the Born diagram (Fig.8.18a). This is expected if hadron production at low Q^2 is dominated by peripheral processes (which may lead to jet-like configurations in the forward direction). Note, however, that also at large jet angles the data lie about a factor of 4 to 5 above the Born prediction. This is not the case at large Q^2: here the jet angular distribution is in agreement with the Born prediction for all jet angles.. (Fig.8.18b). Comparing the events with $\Theta^{Jet} \approx 90°$ for different Q^2 regions one finds that the events at low Q^2 look more spherical than at high Q^2,

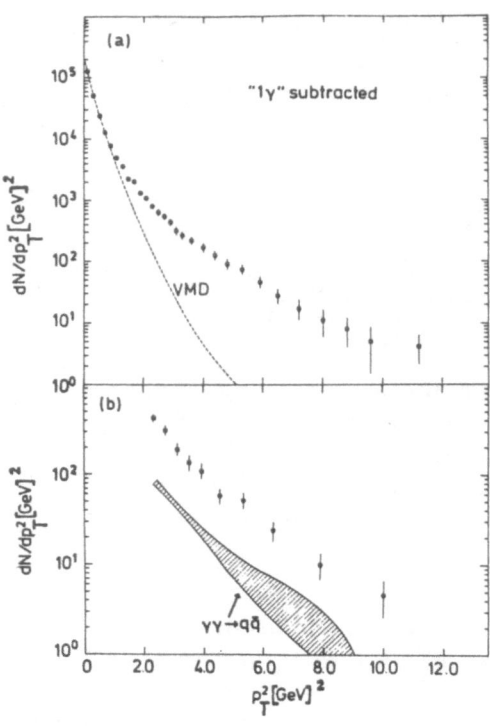

Fig.8.17. Transverse momentum distribution of single charged hadrons for no-tag data (TASSO):

a) dN/dp_T^2 versus p_T^2 distribution of all tracks in an event after subtracting the "1γ" contribution. The dotted line is the contribution expected from VMD.

b) Same as a) but after subtracting the $\gamma\gamma \to \tau^+\tau^-$ and the VMD contributions. The data are compared to the Born approximation for quark pair production represented by a band which shows the maximum variation that can be obtained by changing the fragmentation parameters.

where the event topologies are in perfect agreement with the expectation for a 2-jet event.

In Fig.8.19 the ratio of the measured jet cross section to the Born prediction for fractionally charged quarks is plotted as a function of p_T^{jet} for three different Q^2 regions. For Q^2 below 1 GeV^2 the ratio approaches 1 with increasing p_T^{jet} from above. Remarkably, for Q^2 values above ~ 10 GeV^2, the ratio is consistent with 1 over the whole p_T^{jet} range. Thus the data suggest

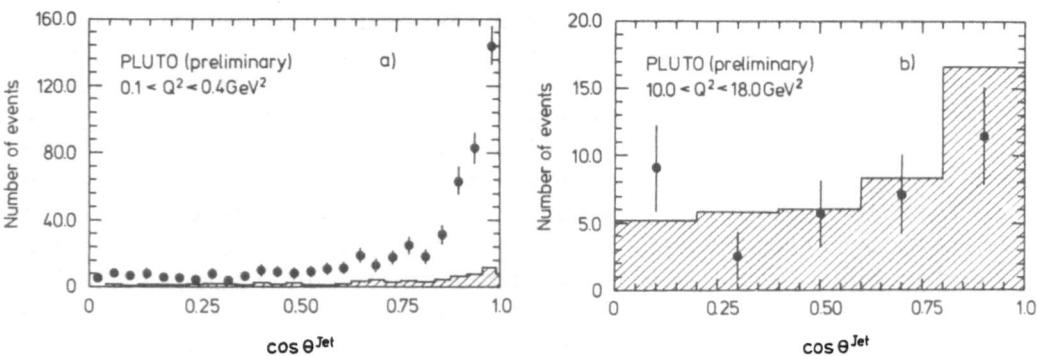

Fig.8.18. Angular distribution of the thrust axis with respect to the $\gamma\gamma$ axis in the $\gamma\gamma$ center of mass system for two different Q^2 ranges (PLUTO). The histograms are the absolute predictions from the quark Born diagram.

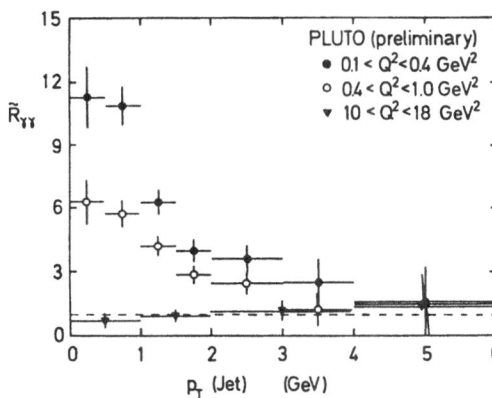

Fig.8.19. Transverse momentum and Q^2 dependence of two-photon jet production (PLUTO): The ratio of the measured jet cross section to the Born prediction for fractional charged quarks is plotted versus the transverse momentum of the jets for three different Q^2 regions.

that in the limit of either large p_T of the jets or large Q^2 of at least one of the photons the data can be described by the lowest order process $\gamma\gamma \to q\bar{q}$ only.

8.4   Discussion of the Results

8.4.1   Comparison to Theoretical Predictions

All experiments studying hard scattering processes in two-photon reactions observe a flattening of the inclusive p_T^2 distribution of single hadrons above $p_T \approx 1$ GeV. In the TASSO single-tag analysis the large-p_T tail exhibits an approximate p_T^{-4} slope (Fig.8.10). This is remarkably different from the p_T behaviour observed in hadronic scattering reactions at similar energies. Such a difference has been predicted to arise from the pointlike component of the photon. However, a quantitative analysis shows, e.g., that for quasi-real photons the data in the range $1.5 < p_T < 3.0$ GeV lie far above the prediction from the Born process $\gamma\gamma \to q\bar{q}$ (Fig.8.17b). This may indicate that in this p_T range higher order processes (multi-jets, higher twist) cannot be neglected.

With increasing Q^2 of one of the photons, the inclusive p_T^2 distribution approaches the Born prediction for all p_T values. A suppression of the hadronic piece of the photon with increasing Q^2 is expected and is consistent with the observed behaviour of the photon structure function (see Chap.9).

Analysing the data in terms of 2–jet configurations, as expected from the Born diagram, the conclusions drawn from the inclusive cross sections are supported. At low Q^2 the cross section is dominated by peripheral processes which apparently also exhibit jet structures but the jets are predominantly produced in the forward direction. At wide angles, however, the excess over the Born contribution seems to be due to events which are more spherical than expected from the Born term. In contrast to the low-Q^2 data, the 2-jet cross section found for large Q^2 is completely consistent with the Born process only. At all p_T^{jet} values the data are consistent with the prediction of the Born term, both in the normalization (using $R_{\gamma\gamma}$ for fractionally charged quarks) and in the event shapes.

In summary, the experimental observations are in agreement with the general expectation that the 0th order QCD process $\gamma\gamma \to q\bar{q} \to 2$ jets becomes dominant for large p_T of the hadrons (or jets) and/or large Q^2 of at least one of the photons. In both limits the value of $R_{\gamma\gamma}$ seems to approach the prediction for fractionally charged quarks.

8.4.2 Integrally versus Fractionally Charged Quarks

The measurement of $R_{\gamma\gamma}$ is supposed to be a clean and theoretically unbiased method to determine the charge of the quarks, in particular to distinguish between the standard fractionally charged quark model (FCQ model) and the integrally charged quark model (ICQ model). The ICQ model has been introduced in Sect.6.3.5. In the following we summarize some of the important points.

In the ICQ model the photon is composed of two pieces, a flavour octet - colour singlet, $(8_f, 1_c)$, and a flavour singlet - colour octet, $(1_f, 8_c)$. Since below colour threshold the observed hadrons have to be in a colour singlet state, the colour octet component of the photon does not contribute to processes involving only one photon. In such processes (e.g. one-photon annihilation) the cross sections obtained with the FCQ and the ICQ model are the same. However, in higher order electromagnetic processes colour octet photons can combine to colour singlets. For example, the product of two colour octets contains a colour singlet (which has to be a flavour singlet) according to the decomposition:

$$(1_f, 8_c) \times (1_f, 8_c) = (1_f, 1_c) + \dots \,. \tag{8.24}$$

Although this decomposition contains a colourless part, the colour octet photons do not contribute to large distance interactions below colour threshold, because in this case the optical theorem would require the existence of coloured physical states. If we consider, however, the process $\gamma\gamma \to q\bar{q}$ only for transverse quark momenta which are large compared to the confinement scale, the colour octet photons contribute fully even below colour threshold. (Note that additional theoretical assumptions are necessary to apply similar arguments to the $\gamma\gamma$ coupling of the η'; see Sect.6.3.5).

Evaluating the Feynman diagrams for $\gamma\gamma \to q\bar{q}$ in the ICQ model (see e.g. /191/), the different colour contributions have to be added coherently and normalized to obtain the required colour singlet projection; the different flavours are added incoherently. With the charge $e_{i\alpha}$ of a quark with flavour i and colour α one obtains for $R_{\gamma\gamma}$, instead of (8.03):

$$R_{\gamma\gamma}(ICQ) = \tfrac{1}{3} \cdot \sum_i (\sum_\alpha e_{i\alpha}^2)^2 = 10/3 \quad \text{for i = u,d,s,c} \quad \text{and} \quad \alpha = r,y,b. \tag{8.25}$$

This expression can be compared to:

$$R_{\gamma\gamma}(\text{FCQ}) = \sum_i \sum_\alpha e_{i\alpha}^4 = 34/27 \quad \text{for } i = u,d,s,c \text{ and } \alpha = r,y,b. \quad (8.26)$$

The value $R_{\gamma\gamma} = 10/3$ is ruled out experimentally, in particular by the large-Q^2 data (Fig.8.19). However, it has been pointed out that a gauge invariant formulation of the ICQ model (gauge-ICQ) /172, 191, 192/ introduces a Q^2 dependence of $R_{\gamma\gamma}$ such that $R_{\gamma\gamma}(\text{ICQ})$ approaches $R_{\gamma\gamma}(\text{FCQ})$ for large Q^2. In the gauge-ICQ model the photons and gluons are gauge bosons arising from the same gauge symmetry. As a consequence photons and gluons can have both electric and colour charge and the gluons are not massless. Since in this model direct couplings between the photon and the gluons can occur, the photon propagator becomes modified in a way that the colour octet component is suppressed for virtual photons. The effective quark charge seen by the photon can be written as (m_g is the gluon mass):

$$Q_{\text{eff}}(Q^2) = Q_{(1c)} + \frac{m_g^2}{m_g^2 + Q^2} \, Q_{(8c)}. \quad (8.27)$$

As a result, only the real photons contribute with the full colour octet component. Another consequence of the direct photon-gluon coupling is an additional contribution to $R_{\gamma\gamma}$, which we refer to as $R_{\gamma\gamma}'$, arising from diagrams with gluons in the final state. The size of the contribution depends on the gluon mass and is comparable to the quark contribution for $m_g \approx 10$ MeV.

The essential predictions of the gauge-ICQ model are /191/:

- $R_{\gamma\gamma}(\text{ICQ}) = 10/3 + R_{\gamma\gamma}'$ for real photons ($Q^2 = 0$).
- $R_{\gamma\gamma}(\text{ICQ}) \to R_{\gamma\gamma}(\text{FCQ})$ for $Q^2 \gg m_g^2$.
- $R_{\gamma\gamma}(\text{ICQ})$ decreases with increasing p_T.
- $R_{\gamma\gamma}(\text{ICQ})$ increases with increasing $W_{\gamma\gamma}$.

The first three predictions are not in obvious disagreement with the data. But that can hardly be used as an argument for the gauge-ICQ model because higher order FCQ processes can produce similar effects. Thus, before one takes the excess of large-p_T events at low Q^2 as evidence for the ICQ model, one has to make sure that the excess events have the proper 2–jet structure and that the jet cross section has the predicted $W_{\gamma\gamma}$ and p_T behaviour.

In conclusion, we want to emphasize that as yet all experimental observations have perfectly reasonable explanations within the standard model of fractionally charged quarks. The data do not (yet?) require any exotic explanation.

9. The Structure Functions of the Photon

In this chapter we discuss the Q^2 dependence of the two-photon total cross section for the case that the Q^2 of one photon is varied and the Q^2 of the other is kept small. For this kinematical situation the cross section is usually expressed in terms of the structure functions of the photon. The scattering of a highly virtual photon on a quasi-real photon target can be thought of as deep-inelastic electron-photon scattering (see Fig.9.1) /193/. The process can be observed at e^+e^- machines by detecting one electron at large angle while the other one is restricted to small angles.

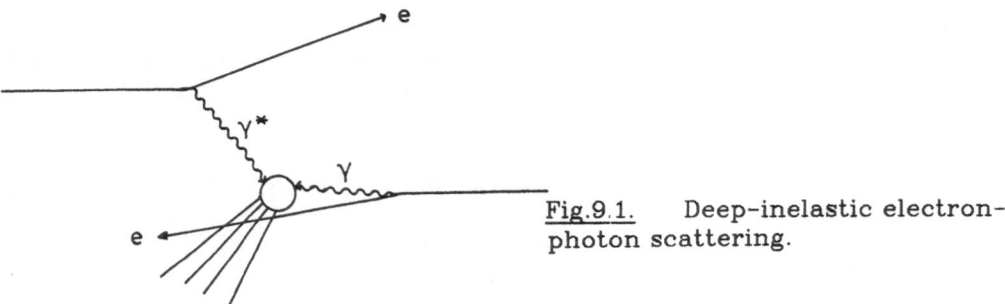

Fig.9.1.   Deep-inelastic electron-photon scattering.

In analogy to deep-inelastic lepton-hadron scattering the virtual photon probes the structure of the target photon. In the VMD picture the analogy to lepton-hadron scattering goes further: The target photon turns into a vector meson and the virtual photon probes a typical hadronic structure as shown in Fig.9.2a. As discussed in Chap.1, the VMD part of the photon is probably a special case of photon splitting into quark pairs. If the photon splits into nearly collinear quarks, the quarks will strongly interact and form vector meson states. On the other hand, if the photon splits into quarks with large transverse momenta the pointlike structure of the photon becomes visible (Fig.9.2b). It has been predicted that the pointlike coupling dominates the structure of the photon at large Q^2 of the probing photon /194/. Most important was the discovery that at large Q^2 the shape and the absolute normalization of the pointlike contribution are calculable to leading order in perturbative QCD /8/. Due to gluon corrections, the QCD result does not coincide with the free parton model prediction even in the asymptotic limit. It has been suggested that a precise measurement of the photon structure functions offers a particularly clean test of QCD and allows the determination of the QCD scale parameter

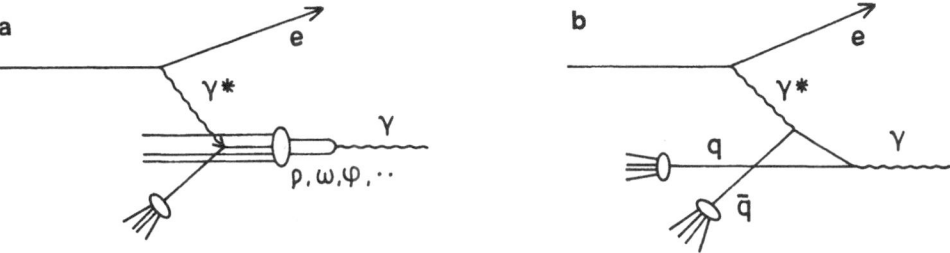

<u>Fig.9.2.</u>   The structure of a photon: a) hadron-like and b) pointlike component of the photon.

Λ. In view of large higher order corrections to the lowest order QCD calculation of the structure functions, the validity of this statement at finite Q^2 has been extensively debated (see e.g. /195, 196, 197/).

The first experimental results on the photon structure functions clearly show the behaviour expected from the dominance of the pointlike coupling of the photon and thus a behaviour which is significantly different from that of hadronic structure functions. The observations are in agreement with the simple assumption that the structure functions can be described as the sum of a pointlike and a hadronic component.

This chapter is organized as follows: after having defined the structure functions, we discuss their meaning for the two limits: hadron-like and pointlike structure. Then we discuss the leading order QCD results and the difficulties arising from higher order corrections. Finally, we present the experimental results together with a discussion of the experimental methods.

9.1   The Definition of the Structure Functions of the Photon

We use the form of the two-photon cross section as given in (2.36) for electron-photon scattering with $q_2^2 \approx 0$:

$$\frac{d\sigma(e\gamma \to eX)}{d\Omega_1 dE_1' d\tilde{\varphi}/2\pi} = \Gamma_T \cdot \left[\sigma_{TT}(W_{\gamma\gamma}, q_1^2) + \varepsilon \cdot \sigma_{LT}(W_{\gamma\gamma}, q_1^2) + \tfrac{1}{2}\varepsilon_1\varepsilon_2 \cdot \tau_{TT}(W_{\gamma\gamma}, q_1^2) \cdot \cos 2\tilde{\varphi} \right].$$

$$(9.01)$$

The following Lorentz scalars are usually introduced:

$$Q^2 = -q_1^2$$
$$P^2 = -q_2^2$$
$$x = Q^2/(2q_1q_2) \approx Q^2/(Q^2 + W_{\gamma\gamma}^2)$$
$$y = y_1 = q_1q_2/(p_1q_2) \approx 1 - (E_1'/E) \cdot \cos^2\Theta_1/2$$
$$\xi = y_2 = q_1q_2/(p_2q_1) \approx 1 - E_2'/E.$$

$$(9.02)$$

The variables used on the right hand side of these formulae are defined in Chap.2 (see also Fig.2.2). The approximations hold for $P^2 \approx 0$. With these variables we get for $P^2 \approx 0$ and $Q^2 >> 4m_e^2$:

$\Gamma_T = \alpha E_1'(2-2y+y^2)/(2\pi^2 \cdot Q^2 \cdot y)$,

and with $f(y) = 2(1-y)/(2-2y+y^2)$

$\varepsilon \approx f(y)$

$\varepsilon_1 \approx f(y)$

$\varepsilon_2 \approx f(\xi)$.   (9.03)

The approximation for ε_2 is valid for small P^2, but $P^2 \gg 4m_e^2$. In terms of the cross sections σ_{TT}, σ_{LT}, $\sigma_{\parallel}$ and $\sigma_{\perp}$, which have been introduced in Chap.2, the structure functions are defined as follows (for different definitions in the literature see e.g. appendix B in /198/):

$F_T(x,Q^2) = \sigma_{TT} \cdot Q^2/(8\pi^2\alpha x)$

$F_L(x,Q^2) = \sigma_{LT} \cdot Q^2/(4\pi^2\alpha)$ (9.04)

$F_X(x,Q^2) = (\sigma_{\parallel} - \sigma_{\perp}) \cdot Q^2/(4\pi^2\alpha)$.

With these structure functions (9.01) becomes ($E_{\gamma 2}$ is the energy of the target photon):

$$\frac{d\sigma}{dxdyd\tilde{\varphi}/2\pi} = \frac{8\pi\alpha^2 E E_{\gamma 2}}{Q^4} [1+(1-y)^2]$$

$$\times [2xF_T(x,Q^2) + \varepsilon F_L(x,Q^2) + \varepsilon_1\varepsilon_2 F_X(x,Q^2)\cos2\tilde{\varphi}]. \tag{9.05}$$

An alternative set of structure functions is defined as:

$F_1(x,Q^2) = F_T(x,Q^2)$

$F_2(x,Q^2) = 2xF_T(x,Q^2)+F_L(x,Q^2)$ (9.06)

$F_3(x,Q^2) = F_X(x,Q^2)$.

With these definitions (9.05) reads:

$$\frac{d\sigma}{dxdyd\tilde{\varphi}/2\pi} = \frac{16\pi\alpha^2 E E_{\gamma 2}}{Q^4} \left[(1-y)F_2(x,Q^2)+xy^2F_1(x,Q^2)+(1-y)\varepsilon_2 F_3(x,Q^2) \cdot \cos2\tilde{\varphi} \right].$$
$$\tag{9.07}$$

The third structure function can only be measured if the angle $\tilde{\varphi}$ between the lepton scattering planes is detected, i.e. if the soft photon is tagged at small angles. Usually the soft photon is not tagged and the F_3 term vanishes. Furthermore, in most experimental situations the $F_1(x,Q^2)$ term in (9.07) can be neglected because y tends to be very small (corresponding to $\varepsilon \approx 1$) so that xy^2 is much smaller than $1-y$. The differential cross section can then be approximated by:

$$\frac{d\sigma}{dx\,dy} \approx \frac{16\pi\alpha^2 E E_{\gamma 2}}{Q^4} (1-y)\ F_2(x,Q^2). \tag{9.08}$$

9.2   The Physical Meaning of the Structure Functions

9.2.1   The Hadronic Component of the Photon

At low Q^2 photons are expected to interact like hadrons. In the VMD picture the target photon turns into a vector meson ρ, ω, φ,... and the virtual photon 'sees' the structure of a hadron (Fig.9.2a). (The VMD part of the virtual, large Q^2, photon is suppressed by the Q^2 dependence of the vector meson propagators.) In such a process the structure function can be interpreted, in analogy to deep-inelastic electron–nucleon scattering, as the sum over the momentum distributions of the partons in a hadron:

$$F_2(x,Q^2)/x \;=\; \sum_i e_i^2 f_i(x); \qquad F_1(x) \;=\; \sum_i e_i^2 f_i(x)/2. \tag{9.09}$$

$f_i(x)$ is the probability to find a parton of type i and charge e_i with a momentum fraction x (in the infinite momentum frame) in the hadron; the sum goes over the partons in the hadron.

Ignoring gluon emission, the x distributions are independent of Q^2 due to the limited transverse momentum of the partons in a hadron (i.e. the structure functions show Bjorken scaling). With gluon radiation and absorption the x distribution shifts to smaller x as Q^2 grows, i.e. as the virtual γ resolves smaller distances, because more momentum goes into radiated gluons (Fig.9.3). In the Bjorken limit ($Q^2 \to \infty$, x fixed) the hadronic structure functions $F_{1,2}^{VMD}$ decrease like $1/\ln Q^2$ for any given x.

We assume that the distributions $f_i(x)$ for the vector mesons are similar to the distributions in a pion, which have been determined from measurements of the Drell-Yan process $\pi^\pm \to \mu^+ \mu^- X$ /199/ to have the form:

$$f_i^\pi(x) \;\sim\; (1-x)/x. \tag{9.10}$$

Taking only the dominant ρ contribution one gets /198/:

$$F_2^{\gamma,VMD}(x) \;\approx\; \alpha\pi/\gamma_\rho^2 \cdot F_2^\pi(x) \;\approx\; \alpha\pi/\gamma_\rho^2 \cdot 0.25 \cdot (1-x). \tag{9.11}$$

The coherent superposition of the lowest vector meson states, ρ, ω, φ, yields for $F_2^{\gamma,VMD}$ /165/:

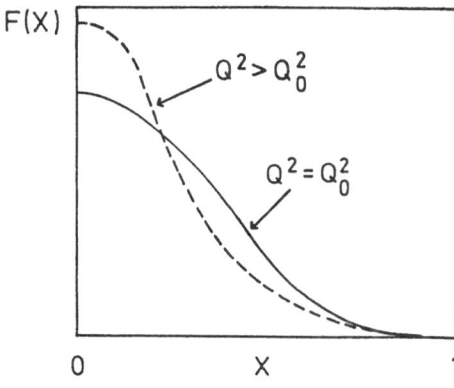

F(X)

$Q^2 > Q_0^2$

$Q^2 = Q_0^2$

0   X   1

<u>Fig.9.3.</u> Q^2 dependence of a typical hadronic structure function as expected from scale breaking QCD effects.

$$F_2^{\gamma,\text{VMD}}(x) \approx \alpha(0.2\pm0.05)(1-x).\tag{9.12}$$

See also the discussion in /200/ where in addition the dependence on the target mass P^2 is included.

9.2.2   The Pointlike Coupling of the Photon

Although the total cross section for two-photon production of hadrons is dominated by VMD-like processes, the pointlike coupling of the photon to quarks is expected to become important in kinematical regions involving high momentum transfers. As an example, we discussed in the last chapter processes with high-p_T particles in the final state. In the case of deep-inelastic electron-photon scattering, the virtual photon has a large momentum transfer to the target photon and hence one might expect that the direct, pointlike coupling of the photon to quarks becomes observable /194/.

Disregarding strong interactions, the pointlike coupling of the photons to quarks is given by the lowest order QED diagram in Fig.9.2b for $\gamma^*\gamma\to q\bar{q}$, i.e. the box diagram (Fig.4.1) contribution to the absorptive part of the $\gamma^*\gamma\to\gamma^*\gamma$ forward scattering amplitude. In this approximation, no unknown hadronic distribution function enters into the structure functions. The basic differences between the VMD picture in Fig.9.2a and the QED graph in Fig.9.2b can be traced back to the different transverse momentum distributions of the partons in the two diagrams. In a hadron the partons have limited transverse momenta. At large Q^2 this leads to a vanishing cross section for longitudinal photons because of helicity conservation at the γ^*qq vertex (Callan-Gross relation). In the diagram of Fig.9.2b in contrast, the quark transverse momenta can be large. As a consequence the longitudinal structure function does not vanish for large Q^2 and the transverse structure function increases logarithmically with Q^2, i.e. there is no scaling even in the quark-parton model. Note, however, that in the asymptotic limit the ratio F_L/F_T vanishes. The non-scaling of the transverse structure function can be demonstrated by calculating the structure functions for the diagram in Fig.9.2b. Neglecting terms $O(m_q^2)$ we obtain (for a quasi-real target photon):

$$\begin{aligned}
F_T(x,Q^2) &= (\alpha\cdot3\cdot\textstyle\sum e_q^4/2\pi)\cdot\left[[x^2+(1-x)^2]\ln(W_{\gamma\gamma}/m_q)^2 + 4x(1-x)-1\right]\\
F_L(x,Q^2) &= (4\alpha\cdot3\cdot\textstyle\sum e_q^4/\pi)\cdot x^2(1-x)\\
F_3(x,Q^2) &= (\alpha\cdot3\cdot\textstyle\sum e_q^4/\pi)\cdot x^3.
\end{aligned}\tag{9.13}$$

Writing

$$\ln(W_{\gamma\gamma}/m_q)^2 = \ln(Q^2/m_q^2) + \ln[(1-x)/x]\tag{9.14}$$

we see that $F_T(x,Q^2)$ is both a function of x and of Q^2 and thus does not

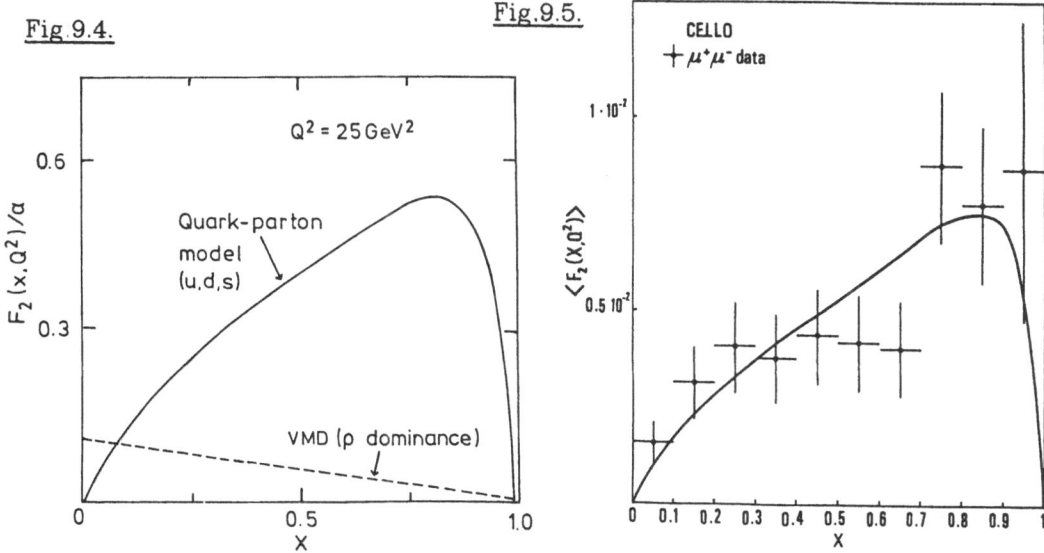

<u>Fig.9.4.</u>   The photon structure function $F_2(x,Q^2)$ predicted by the quark-parton model and the VMD model   (assuming ρ dominance).

<u>Fig.9.5.</u>   The "leptonic" photon structure function determined from the reaction $\gamma\gamma \rightarrow \mu^+\mu^-$ at $<Q^2>=9.5$ GeV^2 (CELLO).   The curve is the QED prediction.

scale. The logarithm is due to the increase of the phase space for transverse momenta with increasing $W_{\gamma\gamma}$.

In Fig.9.4 the expectations from the quark-parton model and from VMD (9.11) are shown. $F_2(x,Q^2)$ is plotted versus x for fixed Q^2. While the hadronic structure function decreases with x, the structure function for a pointlike photon increases with x up to a maximum near x=1. Note that in the quark-parton model the absolute height of the structure function depends on the quark masses (the constituent quark masses give a reasonable description of the data).

In evaluating the diagram in Fig.9.2b a pure QED coupling of the photons to the quarks has been assumed, in complete analogy to lepton pair production by two photons. Lepton pair production at large Q^2 has been analysed by the CELLO group in terms of the structure function $F_2(x,Q^2)$ (Fig.9.5) /47/. In this analysis the CELLO group studied in detail the effects of the various kinematical approximations commonly used in the structure function formalism. They find, e.g., that variations of the target photon mass cannot be neglected if the undetected scattered lepton is not restricted to small angles (anti-tag). To account for this effect, the target photon density has to be expressed as a function of Θ_2 (i.e. the expression (2.35) has to be differentiated with respect to Θ_{2max}) and the structure function in (9.13) has to be modified. The complete expression for the structure function resulting from the Born term can be obtained from the cross sections for lepton pair production (see appendix E in /12/) by using the relations between cross sections and structure functions in (9.04). For small target masses P^2 and light leptons or quarks, with $P^2 \ll Q^2$ and $m^2 \ll W^2$, the following expression for F_2 was derived in /47/ (for quarks F_2 has to be multiplied by $3 \cdot \sum e_q^4$):

$$F_2(x,Q^2,P^2) = \frac{\alpha}{\pi} x \left\{ [x^2 + (1-x)^2] \ln\left(\frac{W^2}{m^2 + P^2 x(1-x)}\right)^2 + 8x(1-x) - 1 - \frac{P^2 x(1-x)}{m^2 + P^2 x(1-x)} \right\}.$$

$$(9.15)$$

For the charm quark and heavier quarks the mass terms in the structure functions cannot be neglected in the presently available Q^2 range. The complete formulae for the structure functions including mass terms can be found in appendix E in /12/. Inclusion of the charm quark results in a typical threshold behaviour as shown in Fig.9.6. The proper treatment of the charm contribution to the structure functions is straightforward in the quark-parton model but not so clear in the framework of perturbative QCD (see below).

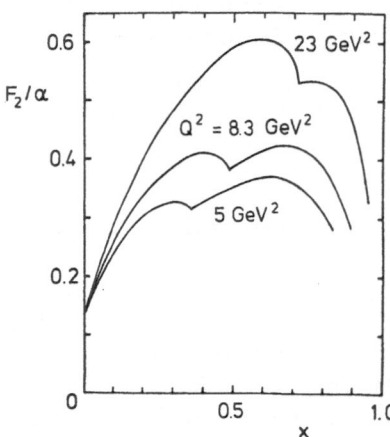

Fig.9.6. Influence of the charm threshold on the structure function (from /201/): For fixed Q^2 the threshold is crossed when going from large x (corresponding to small W) to small x (corresponding to large W). The threshold moves to larger x with increasing Q^2.

9.3   The QCD Evaluation of the Photon Structure Functions

The first QCD evaluation of the structure functions of the photon was done in leading order in the pioneering work of Witten /8/. He found that the shape and the absolute normalization of the structure functions is calculable for large Q^2 in perturbative QCD. This has to be contrasted to the QCD treatment of hadronic structure functions, where only the Q^2 evolution can be calculated. For photons, the leading Q^2 dependence of $F_2(x,Q^2)$ was found to be proportional to $\ln(Q^2/\Lambda^2)$. Hence, the same non-scaling behaviour as in the free quark-parton model is obtained, if one replaces the quark mass by the QCD scale parameter Λ. It was therefore argued that the absolute measurement of $F_2(x,Q^2)$ at large Q^2 provides a straightforward way to determine Λ, if the higher order corrections can be controlled at the finite Q^2's available. For hadronic structure functions the absolute normalization is not calculable and the Λ parameter can only be determined from the Q^2 evolution of the structure function which goes like $1/\ln(Q^2/\Lambda^2)$ for fixed x. It should be noted that for moderate Q^2 such a dependence can be faked by higher twist effects, whereas the increase with Q^2 in the case of the photon structure function cannot.

9.3.1   The Leading Order QCD Result

Witten /8/ analysed the moments of the structure functions

$$M_n(Q^2) = \int_0^1 dx \; x^{n-2} \; F(x,Q^2) \tag{9.16}$$

in terms of the Wilson operator product expansion:

$$M_n(Q^2) = \sum_i C_n^i(Q^2) <\gamma|O_n^i|\gamma>. \tag{9.17}$$

O_n^i are operators of twist two and spin n and the C_n^i are the Wilson coefficients. Higher twist terms are neglected. In contrast to hadronic structure functions not only the quark and gluon operators but also the photon operators have to be included (i.e. i = q, g, γ). It turns out that for a photon the expressions $C_n^i(Q^2)<\gamma|O_n^i|\gamma>$ are of the same order α (the electromagnetic coupling constant) both for the quark and gluon operators as well as for the photon operator. In lowest order QED we have for the photon operator $<\gamma|O_n^\gamma|\gamma> \approx 1$. Since the Wilson coefficients are calculable perturbatively, the whole expression $C_n^\gamma(Q^2)<\gamma| O_n^\gamma|\gamma>$ is calculable. The contributions from the corresponding expressions for quark and gluon operators behave as for hadronic structure functions. Specifically, the photon matrix elements of these operators have non-perturbative contributions. However, Witten found that for large Q^2 the contributions from the photon operators dominate over the hadronic contributions.

Subsequently other authors calculated the structure functions in leading order perturbative QCD with different methods /202, 198/, which all lead to the same result. One of these approaches uses the Altarelli-Parisi equations /203/ to describe the quark and gluon distributions in the photon. In the following we want to sketch the application of this formalism. For this purpose we write the photon structure functions in terms of quark and antiquark distributions $q,\bar{q}$:

$$F_2(x,Q^2)/x = \sum_q e_q^2 \cdot [q(x,Q^2) + \bar{q}(x,Q^2)]. \tag{9.18}$$

The Q^2 evolution of the quark and gluon distributions is given in leading order by the Altarelli-Parisi equation /204/ ($t = \ln(Q^2/\Lambda^2)$):

$$dq(x,t)/dt = \frac{1}{2\pi} \int_0^1 dz \int_0^1 dy \; \delta(x-yz)$$

$$\left\{ \alpha_s(t) \left[P_{qq}(z)\cdot q(y,t) + P_{qg}(z)\cdot g(y,t) \right] + \alpha\cdot P_{q\gamma}(z)\cdot\gamma(y,t) \right\} \tag{9.19}$$

$$dg(x,t)/dt = (\alpha_s(t)/2\pi) \int_0^1 dz \int_0^1 dy \; \delta(x-yz)\left[P_{gq}(z)\cdot\textstyle\sum q(y,t) + P_{gg}(z)\cdot g(y,t) \right].$$

$\alpha_s(t)$ is the running strong coupling constant, α is the fine structure constant, q, g, γ are the quark, gluon and photon distributions as functions of t and x. The $P_{ij}(z)$ are the probabilities that the fraction z of the x value of the particle j goes to the particle species i when t changes from t to t+dt.

The term containing the photon distribution, being of order α, is negligible for a hadron target, whereas for a photon target the gluon and quark distributions are also of order α and hence all three distributions are comparable. In leading order QED the photon density is just $\gamma(y,t) = \delta(y-1)$ and $P_{q\gamma}$ can be approximated by the quark–parton model result (9.13):

$$P_{q\gamma}(x,t) = dq^{box}(x,t)/dt = (3\alpha\cdot e_q^2/2\pi)\cdot[x^2+(1-x)^2]. \tag{9.20}$$

Thus the photon contribution leads to an inhomogeneous term in the Altarelli–Parisi equation (9.19) which is not present for hadron targets.

The solution of (9.19) for the structure functions has the form:

$$F_T^{LO}(x,Q^2) = (\alpha/2\pi)[a(x) + b(x)\cdot\ln(Q^2/\Lambda^2)] + f_T^h(x,Q^2)$$
$$F_L^{LO}(x,Q^2) = (\alpha/2\pi)\cdot c(x) + f_L^h(x,Q^2) \tag{9.21}$$
$$F_3^{LO}(x,Q^2) = F_3^{BOX}(x,Q^2).$$

For large Q^2 $f_T^h(x,Q^2)$ and $f_L^h(x,Q^2)$ vanish as $1/\ln Q^2$, i.e. like a hadronic structure function. In leading order the functions a(x), b(x) and c(x) are independent of Q^2 and calculable. Thus, for large Q^2, the structure functions have the same Q^2 dependence as in the parton model and an x dependence which is modified in a calculable way:

$$F_T^{LO}(x,Q^2) = (\alpha/2\pi)[a(x) + b(x)\cdot\ln(Q^2/\Lambda^2)]$$
$$F_L^{LO}(x,Q^2) = (\alpha/2\pi)\cdot c(x) \tag{9.22}$$
$$F_3^{LO}(x,Q^2) = F_3^{BOX}(x,Q^2).$$

Retaining only the leading logarithm, we have in addition:

$$F_2^{LO}(x,Q^2) \approx F_T^{LO}(x,Q^2). \tag{9.23}$$

The $\ln(Q^2/\Lambda^2)$ dependence of $F_T(x,Q^2)$ arises from an integration of the quark propagator $\sim 1/(p_T^2+m_q^2)$ in Fig.9.2b over the available phase space for p_T (see e.g. /165/). For light quarks the lower boundary of integration is given by Λ^2, because the kinematical regime $p_T^2<\Lambda^2$ is governed by nonperturbative effects. The nonperturbative part is usually accounted for by adding a hadronic (VMD) contribution to the structure function. The pointlike part (corresponding to the integral over $\Lambda^2 < p_T^2 < Q^2$) is perturbatively calculable and is expected to dominate at large Q^2 allowing the determination of Λ from the absolute normalization of $F_T(x,Q^2)$. The large x region should be best suited to measure Λ, because the hadronic contribution becomes small with increasing x. However, close to x=1 uncertainties due to threshold effects have to be expected. At fixed Q^2 large x corresponds to small $W_{\gamma\gamma}$, thus above some x value mass terms can no longer be neglected.

At moderate Q^2 the neglect of mass terms is certainly not correct for charmed and heavier quarks, even at medium x values (see Fig.9.6). Since explicit QCD calculations for the charm quark are not yet available, comparison of theory and experiment is usually done by combining the quark-parton model result for the charm quark with the QCD result for the light quarks.

9.3.2   Higher Order QCD Corrections

The determination of the QCD parameter Λ from a measurement of the structure functions requires the knowledge of the higher order corrections. The size of these corrections provides an estimate for the convergence of the perturbation expansion. The neglect of other than the leading $\ln(Q^2/\Lambda^2)$ terms corresponds to a redefinition of Λ and thus Λ is not unambiguously defined. In particular, a direct comparison to Λ values obtained in other processes is not meaningful.

The authors of /205/ extended Witten's calculations to include next-to-leading order contributions. They find large corrections especially in the low and high x region. The large corrections at high x values are worrisome because this is the region where the QCD predictions were supposed to be safest. However, it has been pointed out /206/ that these large corrections are at least partially caused by the wrong treatment of kinematical thresholds for x→1 in lowest order. For fixed Q^2, as x approaches 1, $W_{\gamma\gamma}$ goes to 0 crossing the threshold for hadron production. The effect of the leading logarithm approximation can be demonstrated best with the quark-parton result (9.13) in which the logarithmic dependence of the transverse structure function is given by (9.14):

$$\ln(W_{\gamma\gamma}/m_q)^2 = \ln(Q^2/m_q^2) + \ln[(1-x)/x].$$

In the leading logarithm approximation the term $\ln[(1-x)/x]$ is neglected. In Fig.9.7 the exact formula is compared to the leading logarithmic approximation for three different Q^2 values (5, 25, 100 GeV²). Above x≈0.8 the approximation is bad. The leading order QCD result, where $\ln(Q^2/m_q^2)$ is replaced by $\ln(Q^2/\Lambda^2)$, probably suffers from similar problems close to x=1 and at finite Q^2 /206/. It has been suggested that the argument of the

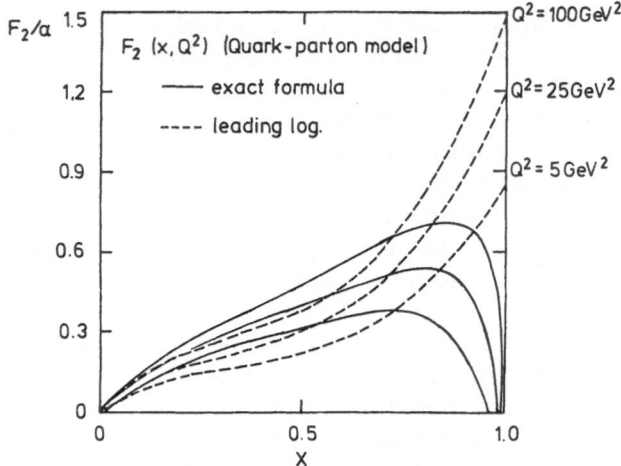

<u>Fig.9.7.</u>   The photon structure function F_2 in the quark-parton model: The full curve is the exact formula and the dashed curve the leading logarithm approximation.

logarithm be modified to improve the convergence properties of the perturbation series /195, 196, 197, 198/.

The large corrections at small x are not unexpected, because in this region the hadronic component of the photon dominates (which cannot be calculated perturbatively). At least at low x it seems to be impossible to disentangle the hard perturbative component from the hadronic, VMD-type component /207, 208/. A careful study of the structure of the QCD results even suggests /197, 209, 210/ that the commonly used procedure of adding independently the pointlike and the hadronic piece (approximated by VMD) has no justification at any x value. It has been found that the part in the QCD result which can be associated with the pointlike piece develops singularities at x=0 at least as bad as $x^{-5.33}$. The occurrence of such singularities, which seem to get even worse with increasing order in α_s, means that the QCD calculations are not safe at any x value. The origin of the singularities has been explained /195/ as a result of infrared divergencies in the gluon radiation corrections, which are characteristic for any bremsstrahlung process (see Fig.9.8a). One expects that such singularities cancel when adding virtual corrections, such as the diagram in Fig.9.8b. The quark-quark interaction represented by this diagram, however, is usually associated with the hadronic, VMD-like piece of the photon. On the other hand, the simple VMD estimate cannot reproduce the singularity structure around x=0 which is needed to cancel the singularities arising from the pointlike piece. Thus, at this point, it appears that the hadronic and the pointlike piece of the photon cannot be separated in a simple way. Consequently, the derivation of the QCD parameter Λ from the absolute normalization of the transverse structure function is questionable.

<u>Fig.9.8.</u>   Examples for gluon corrections at the $\gamma q \bar{q}$ vertex.

This problem has been extensively discussed by many authors /195, 196, 197, 208, 210/. It may turn out that the situation is similar to deep-inelastic lepton-hadron scattering, where Λ can only be determined from the Q^2 evolution of the structure function. Accordingly, a solution of the QCD equations for the photon structure function would require the knowledge of the parton distributions at some finite value Q_0^2 (instead of the distributions at asymptotic Q^2) /210/. The distributions at Q_0^2 have to be determined experimentally. Another possible approach, which leads as well to a parameter dependence in the Λ determination, has been proposed in /208/: The authors developed a procedure to regularize the singularities at x=0 which requires an additional parameter to be determined from the data.

Some theoreticians consider such approaches to be too pessimistic and regard the whole affair as a temporary technical problem /211/. After all, they argue, the measurements of the structure function show that corrections to the asymptotic result must in fact be small in a rather wide range of medium x values, suggesting that there might be a clever way to overcome the problems.

A source of confidence for the optimists is the fact that the large higher order corrections at low x are absent in the case of highly virtual target photons $(P^2>>\Lambda^2)$ /209, 212, 213/. However, in this case the sensitivity to the value of Λ is lost, because the scale breaking effect is then given by a term $\sim\ln(Q^2/P^2)$. QCD calculations for $\Lambda^2<<P^2<<Q^2$ have been carried out in /209, 213/. This kinematical range was found to be free of the difficulties encountered at $P^2\approx0$ and the authors emphasize that this region is particularly suited for clean QCD tests. Thus, even if the Λ determination from the photon structure function turns out to be not easier or even more difficult than in the hadronic case, the special nature of the photon offers a lot more possibilities to study QCD.

9.4 Experimental Results on the Photon Structure Function

The first analysis of deep-inelastic electron-photon scattering was done by the PLUTO group at momentum transfers $1<Q^2<15$ GeV^2 /214/. This analysis showed very clearly that the hadronic behaviour of the photon, which totally dominates the low Q^2 and low p_T regime, is indeed suppressed at high Q^2 in favour of the pointlike component of the photon. With results on the photon structure function for $10<Q^2<60$ GeV^2, the JADE collaboration entered a domain where the determination of the QCD parameter Λ may become possible /215/. A paper by the CELLO group /216/ and a recent preliminary analysis by the JADE group /165/ show that, at presently available energies, the structure functions can be measured up to $Q^2 \approx 300$ GeV² by detecting the scattered electron in the central detector /216/.

9.4.1 The Analysis Methods

The procedure of determining structure functions is in principle the same as the one discussed for the determination of the total hadronic cross section of two quasi-real photons. The main difference is the larger Q^2 of one of the photons. Since in general the single-tag condition is applied, the mass of the hadronic system has to be inferred from the measured final state particles. Thus, in principle, the same problems are expected from the model dependence of the acceptance etc., as discussed for the total cross section analyses in Chap.7. But it turns out that the kinematics are more favourable if one photon is tagged: The hadronic system has to balance the transverse momentum of the scattered lepton thus obtaining an appreciable boost towards the central detector which in turn increases the trigger and particle detection efficiency. For not too large tag angles Θ the transverse momentum of the hadronic system is approximately given by $p_T^{had} \approx Q^2/(E \cdot \Theta)$. For instance, at beam energies of about 15 GeV and $Q^2 > 10$ GeV² one has $p_T^{had} > 3$ GeV. Due to this transverse boost the measured hadronic mass W_{vis} is much closer to the true mass $W_{\gamma\gamma}$ than in the case of the total cross section measurements. In the JADE case, for example, the relation is on the average: $\langle W_{vis} \rangle \approx 0.7 \cdot \langle W_{\gamma\gamma} \rangle$. For the evaluation of the structure functions the variable x_{vis} is introduced, which is the variable $x = Q^2/(Q^2 + W_{\gamma\gamma}^2)$ as obtained from the measured values of Q^2 and $W_{\gamma\gamma}$ (= W_{vis}). Since W_{vis} in general is smaller than $W_{\gamma\gamma}$, x_{vis} tends to be larger than x. In Fig.9.9 we show the x versus x_{vis} relation for the JADE analysis /217/. The bars indicate the spread in x (x_{vis}) for the case of

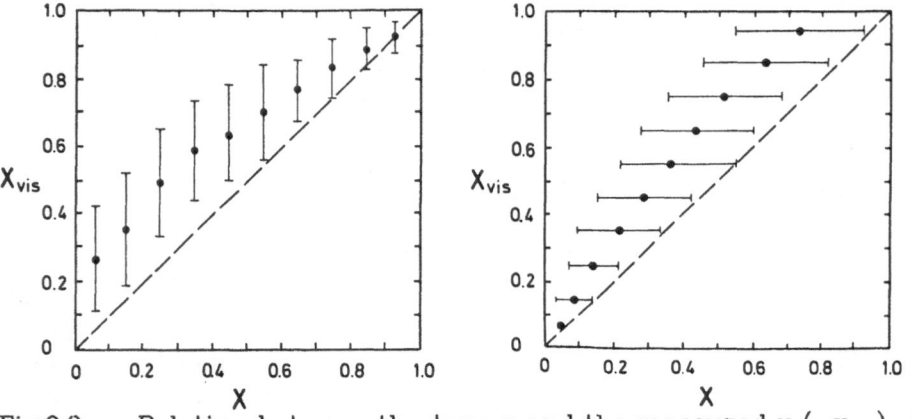

Fig.9.9.   Relation between the true x and the measured x (=x_{vis}) for the JADE detector.

fixed x_{vis} (x). The shift of x_{vis} relative to x amounts to about 0.1 to 0.2 units.

First preliminary results from the improved PLUTO detector ("PLUTO 81") show that the addition of a forward spectrometer (covering a polar angular range down to 5°) improves the $W_{\gamma\gamma}$ determination significantly. In Fig.9.10 the W_{vis}/W distributions obtained with and without the forward spectrometer are compared. The average W_{vis}/W moved up from ~0.6 to ~0.8 and the distribution became much narrower. With the new detector, the PLUTO group has demonstrated that a reliable unfolding of the x distribution is possible and nearly independent of the fragmentation model used to describe the hadrons /165/. A numerical procedure for unfolding the x distribution was presented in /218/.

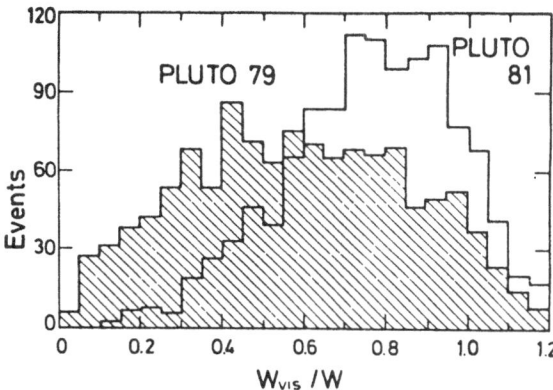

<u>Fig.9.10.</u>   Distribution of the observed fraction of the true $\gamma\gamma$ invariant mass for the PLUTO detector. Open histogram: with forward spectrometer ("PLUTO 81"); shaded histogram: without forward spectrometer (before 81).

What kind of <u>model for the final state hadrons</u> should be used to calculate the detector acceptance by a Monte Carlo simulation? Different processes may contribute at high Q^2 ranging from purely hadronic (VMD-type) scattering with two jets along the $\gamma\gamma$ direction to hard scattering processes, like $\gamma\gamma \to q\bar{q}$, leading to two quark jets with an angular distribution given by QED. At high Q^2 and not too small x the latter process is expected to dominate. In most analyses models have been used where quark pairs are generated like two-photon produced μ pairs with subsequent fragmentation into hadron jets. The PLUTO group found that the other extreme assumption, namely a VMD-type model as used in their analysis of the total cross section, does not lead to significantly different results /214/.

In most of the analyses which will be discussed below, the x distribution has not been unfolded so that a <u>comparison to theoretical predictions</u> is not directly possible. In these cases one has to simulate x_{vis} distributions for the different theoretical models, which then can be compared to the measured x_{vis} distribution. From the discussion of the problems of higher order corrections it is clear that the comparisons with QCD predictions and especially the attempts to determine the QCD parameter Λ should be taken with a grain of salt. The experimenters compare their data to the quark-parton model and to leading order or higher order QCD results, with or without adding a VMD component to describe the hadronic piece of the photon (disregarding for the practical application the problems inherent to this procedure). The c quark contribution, which becomes very important at high Q^2 (see Fig.9.6), has not been explicitly included in the QCD calculations. Usually, the parton model result is used for the c quark contribution.

It has been pointed out that the target mass effect cannot be neglected if the target photon is not restricted by "anti-tagging" to very small Q^2 values /165/. For large target photon masses P^2 the leading scale breaking term becomes $\ln(Q^2/P^2)$ instead of $\ln(Q^2/\Lambda^2)$. The following modification was suggested if P^2 is comparable to Λ^2 /165/:

$$\ln(Q^2/\Lambda^2) \rightarrow \ln(Q^2/(\Lambda^2+P^2)), \tag{9.24}$$

or in analogy to the corresponding QED formula:

$$\ln(Q^2/\Lambda^2) \rightarrow \ln(Q^2/(\Lambda^2+x^2P^2)). \tag{9.25}$$

Using this correction for target mass effects in experiments without an anti-tag requirement, the determined Λ value can change by as much as 50%.

The main sources of experimental background to deep-inelastic electron-photon scattering come from

 —   two-photon QED reactions (especially τ pair production)
 —   one-photon annihilation processes
 —   inelastic Compton scattering.

The $\gamma\gamma$ QED processes can be suppressed by multiplicity cuts and by using particle identification. One-photon annihilation events can contribute if final state hadrons simulate a tag or if a radiated hard photon converts and thus simulates a tag. The first case is rare because the hadrons usually do not deposit enough energy in the forward shower counters. A cut in the energy required for a tag (typically $E_{tag}>0.40 \cdot E$) reduces also the contributions from converted radiated photons. The inelastic Compton scattering process (diagrams b and c in Fig.2.1) is also suppressed by a cut in the tag energy, because such a cut removes backward scattered electrons which are enhanced in the Compton process (u-channel pole). The forward scattered electrons from the Compton process become an increasingly important background for larger tag angles (i.e. large Q^2) and smaller $W_{\gamma\gamma}$ (i.e. larger x). In the JADE experiment at $<Q^2> \approx 23$ GeV^2, for example, the inelastic Compton contribution above x~0.8 reaches about the same magnitude as the signal, i.e. electron-photon scattering /217/ (see also /198/).

The background processes listed above are in principle calculable and can be subtracted using Monte Carlo simulations. Note however, that the calculations of the radiative one-photon process and of the inelastic Compton process require the knowledge of the one-photon annihilation cross section at low energies, which may introduce some uncertainties.

9.4.2   The PLUTO Measurement

The first measurement of the photon structure function has been reported by the PLUTO group /214/. The experiment covered the Q^2 range from 1 to 15 GeV² , as determined by detecting one of the scattered leptons in an angular range from 100 to 250 mrad for beam energies around 15.5 GeV. The reconstructed invariant mass of the hadrons, W_{vis}, was required to be larger than 0.75 GeV, corresponding roughly to more than 1 GeV for the true $W_{\gamma\gamma}$[9]. For an integrated luminosity of 2.5 pb⁻¹ , 117 events with an estimated background of 6 events (mainly beam-gas scattering) were found. The inelastic Compton background is negligible for the small electron scattering angles and the required tag energy of more than 8 GeV.

In Fig.9.11 the Q^2 dependence of the cross section $\sigma_{TT} + \varepsilon\,\sigma_{LT}$ is plotted including points from the low Q^2 analysis (Chap.7). The curve in this plot shows how the cross section would drop if the coupling of the virtual γ were dominated by the ρ pole. We see that the data are in good agreement

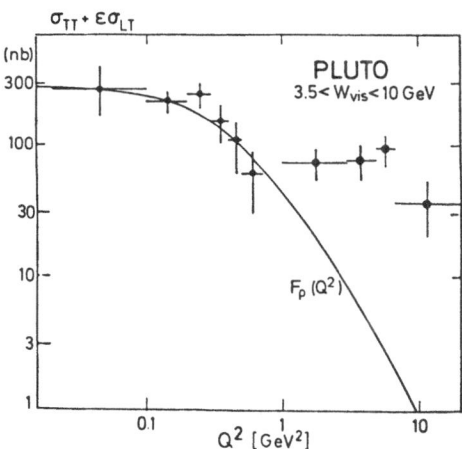

<u>Fig.9.11.</u> Q^2 dependence of the total cross section measured with the PLUTO detector as operated until 1979. The data are compared to the VMD prediction for ρ pole dominance.

---

[9]   In deep-inelastic eN scattering scaling, as expected from the parton model, is observed down to the resonance region if one averages the cross section over the resonances. One might argue that also in the case of electron-photon scattering the parton picture is applicable at low $W_{\gamma\gamma}$.

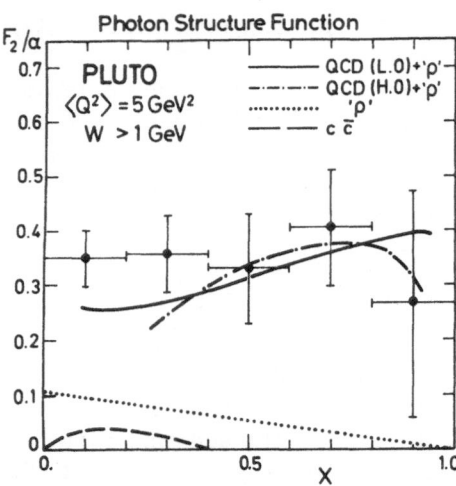

Fig.9.12. F_2/α versus x for $<Q^2>=5$ GeV^2 measured with the PLUTO detector as operated until 1979. The curves are explained in the text.

with this model up to $Q^2 \sim 1$ GeV^2. For $Q^2 > 1$ GeV^2 a larger cross section is observed indicating contributions from hard scattering processes. Such a deviation of the cross section from the VMD prediction is expected if the the virtual photon acts at large Q^2 as a pointlike probe of the target (that is not different to deep-inelastic lepton-nucleon scattering).

The evidence that not only the probe but also the target photon shows the pointlike coupling to the constituents of matter is obtained from a study of the structure functions. In Fig.9.12 F_2/α is plotted versus the true x, i.e. the plot shows the x dependence as unfolded from the x_{vis} distribution. The structure function looks flat with an average value of about 0.35, strikingly different from the expectation for a purely hadronic behaviour of the photon. The latter would give a falling structure function according to the VMD picture (the dotted curve shows the prediction for ρ dominance). The leading order QCD prediction (LO) for light quarks was taken from /219/ with $\Lambda_{LO} = 0.2$ GeV; the curve including next-to-leading order corrections (HO) was calculated in the $\overline{MS}$ scheme using $\Lambda_{\overline{MS}} = 0.2$ GeV. For the covered Q^2 range the charm quark contribution (dashed curve) is small and restricted to small x values.

We conclude that the measured structure function at $<Q^2> \sim 5$ GeV^2 is indeed already dominated by the elementary coupling of the photon and that the agreement with QCD calculations is good, though the simple quark-parton model gives an equally good description.

A good description of the Q^2 dependence of two-photon reactions has also been obtained using the "extended vector meson dominance" model (EVDM) /220/. In this model the photon is represented by an infinite number of vector mesons. With properly adjusted parameters and using information from other reactions involving photons, the Q^2 dependence of the total cross section in Fig.9.11 as well as the structure function in Fig.9.12 are described. The EVDM is able to simulate both the hadronic and the pointlike component of the structure function. A smooth transition between both components is an intrinsic feature of this model. However, the

EVDM is just a phenomenological description of the data, and in particular, it is not meant to be an alternative to QCD.

9.4.3 The JADE Measurement

The JADE group has carried out an analysis of the photon structure function at much higher Q^2 values, $10<Q^2<60$ GeV^2 with $<Q^2>=23$ GeV^2 /215/. They have used the lead glass endcap shower counters of the JADE detector (245 to 500 mrad) as tagging devices. The energy of the scattered lepton had to exceed 50% of the beam energy (the average beam energy was 16.8 GeV). W_{vis} has been determined from the charged tracks measured in the central drift chamber, which covers 97% of the solid angle and from the shower energy in the surrounding lead glass shower counters which cover 90% of the solid angle. At least three charged tracks or two and a photon shower have been demanded in addition to the tag. To remove one-photon annihilation events with a radiated hard photon faking a tag, the events have been accepted only if the resulting longitudinal momentum of the whole detected system (including the detected electron) has been larger than 7.5 GeV. This takes advantage of the fact that $\gamma\gamma$ events usually loose one high energy electron in the beam pipe. Events with converted photons faking a tag have also been removed by requiring that the dE/dx value measured in the drift chamber corresponds to that for a single electron. For an integrated luminosity of 20.2 pb^{-1} 176 events survived the cuts.

The statistical subtraction of various background contributions determined by Monte Carlo simulation reduces this number to 125 ± 15 deep-inelastic electron-photon scattering events. The most severe background comes from the reaction $e^+e^-\to e^+e^-\tau^+\tau^-$ (25 ± 2.5 events) and from hadronic one-photon annihilation (14 ± 5.5 events). The other background sources, $e^+e^-\to\tau^+\tau^-$, inelastic Compton scattering and beam-gas scattering, contribute about 4 events each.

The JADE group did not make an attempt to unfold the x dependence of the structure function from the measured x_{vis} distribution. The correlation between x and x_{vis} has been shown in Fig.9.9. x_{vis} is determined from the measured W_{vis}, which is on average 30% smaller than the true $W_{\gamma\gamma}$. For comparison with the data theoretical distributions have been obtained via Monte Carlo simulation. Three different theoretical predictions for $F_2(x,Q^2)$ have been considered (the F_1 contribution to the cross section is small compared to the F_2 contribution, see Sect.9.1):

1. The quark-parton model (QPM), (9.13). For the c quark the exact formula, without neglecting nonleading mass terms, was used. The masses of the u, d, s, c quarks were assumed to be 0.3, 0.3, 0.5, 1.6 GeV, respectively.
2. The leading order QCD calculations (LOQCD) /8/, given in (9.22). The c quarks were treated like the light quarks with the exception that the c quark threshold was taken into account by setting the charm part of

$F_2(x,Q^2)$ to zero for $x>x_{thr}=Q^2/(Q^2+4m_c^2)$. Alternatively, a combination of the QCD prediction for the light quarks and of the quark-parton model for the charmed quark was used.

3.   The QCD calculations including higher order corrections (HOQCD). The calculations were taken from /213/, where the QCD parameter Λ is defined in the $\overline{MS}$ renormalization scheme.

In all cases the hadronic part of the structure function was neglected. For the construction of the hadronic final state, quark pairs were generated according to the pure QED process $e^+e^-\to e^+e^-q\bar{q}$ /38/ with subsequent fragmentation of the quarks into hadrons à la Field-Feynman /187/. It should be remarked that the angular distribution for $\gamma\gamma\to q\bar{q}$, taken as a QED process, depends on the mass of the quarks. Possible QCD modifications of the angular distribution have not yet been investigated.

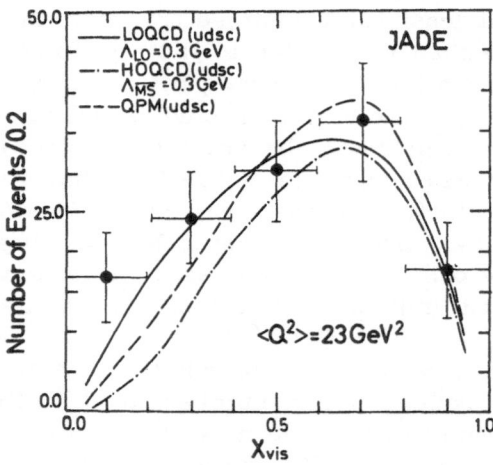

Fig.9.13.   Uncorrected x distribution at $<Q^2>=23$ GeV^2 (JADE). The data are compared to Monte Carlo predictions which are described in the text.

In Fig.9.13 we see that the QCD predictions with adjusted Λ parameters are in good agreement with the data. However, the quark-parton model describes the data as well. Despite the as yet unresolved theoretical problems concerning the determination of Λ (see above), it is useful to test the sensitivity of the data to changes of the Λ parameter in the different theoretical models. Table 9.1 contains the Λ parameters determined from

Table 9.1.   Determination of the Λ parameter from the photon structure function (JADE): Λ is obtained using the lowest order formula (Λ_{LO}) and including higher order corrections in the $\overline{MS}$ scheme ($\Lambda_{\overline{MS}}$). The c quark contribution is treated in three different ways.

Model	Λ_{LO} [GeV]	$\Lambda_{\overline{MS}}$ [GeV]
QCD(udsc)	0.28 +0.13 −0.09	0.22 +0.10 −0.07
QCD(uds) + QPM(c)	0.21 +0.17 −0.09	0.18 +0.12 −0.07
QCD(uds)	0.07 +0.05 −0.03	0.06 +0.05 −0.03

x_{vis}>0.4 for the different assumptions. The range of Λ values obtained, $\Lambda_{LO}\approx\Lambda_{\overline{MS}}$=0.1...0.35 GeV, is consistent with those obtained in deep-inelastic lepton-nucleon scattering.

9.4.4 Most Recent Results

At the 1983 $\gamma\gamma$ workshop in Aachen new data on the structure functions of the photon were presented /165, 221/. Most of the results are still preliminary, but they illuminate clearly what the future developments in this field will be. As can be seen from Table 9.2 the experiments now cover a quite impressive Q^2 range, up to $Q^2 \approx 300$ GeV^2.

Table 9.2. Q^2 ranges covered by different experiments

Experiment	Q^2 range [GeV]	$<Q^2>$ [GeV]	Events
CELLO /216/	30 – 300	100	18
CELLO /221/	3 – 25	9	215
JADE /215/	10 – 55	24	400
JADE /165/	40 – 220	100	25
MAC /165/		32	150
PLUTO /165/	1 – 18	6	2000
PLUTO /165/	15 – 150	45	120
TASSO /165/	10 – 50	23	200

A large fraction of the new results came from the PLUTO group, who presented data taken with the new detector specialized for two-photon physics ("PLUTO 81"). The higher statistics data now available allow a more detailed analysis and the extension of the measurements to kinematical regimes with small cross sections. The first measurement of the photon structure in a double-tag experiment ($Q^2\approx5$ GeV^2, $P^2\approx0.4$ GeV^2) by the PLUTO group deserves to be especially mentioned. This is a first approach towards the ultimate goal to determine the complete set of structure functions in double-tag experiments where the kinematics of both photons is measured. It has been emphasized that the kinematical region $\Lambda^2<<P^2<<Q^2$ is less affected by higher order QCD corrections and thus particularly well suited for clean QCD tests /200/. The preliminary result of the PLUTO group (Fig.9.14) is in good agreement with the QCD prediction of /200/.

Measurements at the highest Q^2 values accessible at PETRA and PEP, up to about 300 GeV^2, have been performed by the CELLO /216/ and JADE /165/ groups using the central detectors for tagging. Figure 9.15 shows the x_{vis} distribution obtained by JADE for $<Q^2>$ = 100 GeV^2.

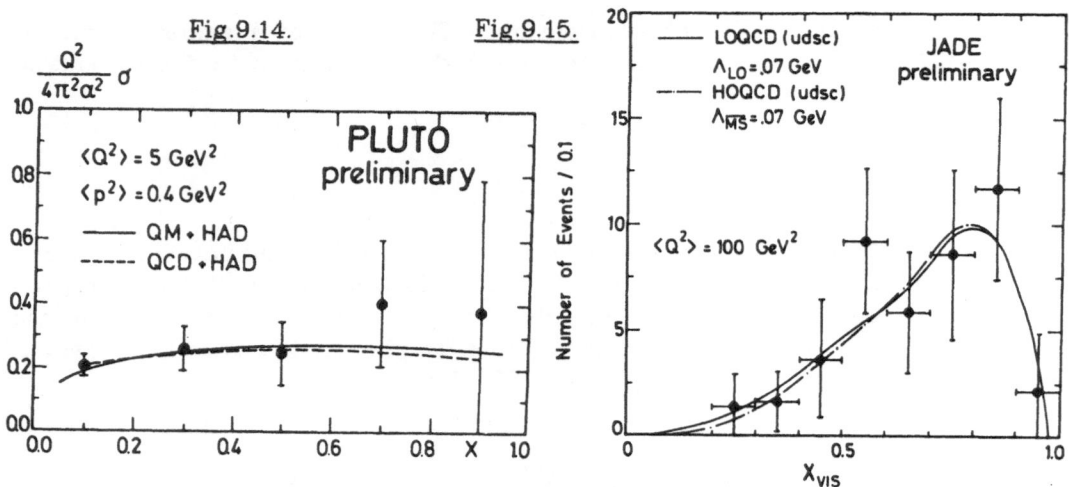

Fig.9.14.   Virtual photon structure function measured in the double-tag mode as a function of x averaged over Q^2 and P^2 (PLUTO, detector 1981-82). The data are compared to the prediction of the quark-parton model (full curve) and to the QCD prediction (dashed curve). In both cases a hadronic piece (HAD) estimated by VMD is added.

Fig.9.15.   Uncorrected x distribution for $\langle Q^2 \rangle$ = 100 GeV^2 compared to lowest order and higher order QCD calculations (JADE).

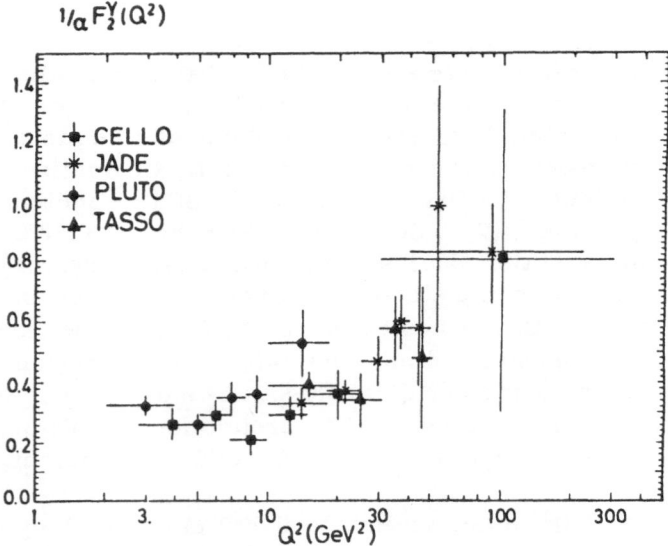

Fig.9.16. Q^2 dependence of $F_2(x,Q^2)$ (compiled data from several experiments).

The QCD predictions for the Q^2 dependence of the structure functions are right now on much safer ground than the predictions for the x dependence and for the absolute normalization. As discussed in Sect.9.3, it may well be that only the Q^2 dependence of the transverse structure function allows a determination of the Λ parameter. Figure 9.16 shows the Q^2 dependence of $F_2(x,Q^2)$ as obtained by combining the (mostly preliminary) results of many experiments. The data are averaged over a medium x range (which is however somewhat different for the different experiments). The plot shows clearly the expected rise of $F_2(x,Q^2)$ with Q^2. However, a more careful study is needed to see how much of the rise with Q^2 can be attributed to the logarithmic scale breaking predicted by QCD. A Q^2 dependence can result e.g. from the finite size of the x intervals via the x-Q^2 correlations or by the onset of charm production. In /165/ it was suggested that the Q^2 dependence of F_2 should be presented for small x intervals and the charm contribution, which is not sensitive to the Λ parameter, should be subtracted. This would allow a direct comparison with the QCD prediction for light quarks.

9.5 Concluding Remarks

The photon structure function F_2 is found to behave differently from the structure function of a hadron. That is interpreted as being due to the pointlike coupling of photons to quarks. All measurements can be well described by a sum of a pointlike and a hadronic component of the photon. The experiments are not yet able to distinguish between a pointlike part calculated in the free quark-parton model and the QCD evaluation of this component. The chances to measure genuine QCD effects in the photon structure functions and to determine reliably the QCD parameter Λ depend mainly on future improvements in understanding the higher order QCD corrections.

10. Summary

In this review, we have described the two-photon scattering experiments at e^+e^- storage rings. In these experiments hadron dynamics has been studied over a wide kinematical range: from resonance physics at low energies to high-p_T physics and the study of deep-inelastic scattering processes.

A large amount of information has been collected in the resonance region. The measurements of two-photon couplings are an important contribution to our understanding of hadronic bound states. In particular, they provide critical tests for models which include multi-quark states and/or gluonium states.

The interaction of photons with hadrons has been investigated in two-photon scattering reactions under various conditions. Quasi-real photons in interactions with low momentum transfer seem to behave like hadrons. On the other hand, the first results on two-photon hard scattering processes and on the structure functions of the photon confirm that the photon has also a pointlike coupling to the constituents of matter. In these reactions QCD predictions have been tested by measuring cross sections and event topologies.

Higher energies, such as will be obtained at the planned storage rings TRISTAN in Japan and LEP in Europe, would be advantageous for the study of hard processes and structure functions (cleaner jet topologies, extended Q^2 range). On the other hand, for resonance physics (below ~3 GeV) the energies available at PEP and PETRA are sufficient. Some experiments may even be easier to do at lower beam energies (e.g. double tag experiments to measure the total cross section).

In the following we summarize the achievements of the two-photon experiments and point out what remains to be done, naturally with an unavoidable personal bias.

QED REACTIONS have been tested in a wide energy and Q^2 range and were found to agree with the theoretical predictions to within 10%. More sensitive tests require the inclusion of radiative corrections in the comparison of the data to theory.

Elastic photon-photon scattering has as yet only been observed in a resonant process ($\gamma\gamma \to \eta \to \gamma\gamma$). The actual observation of light-light scattering would be a nice physics result.

NON-RESONANT HADRON PAIR PRODUCTION allows a test of low energy theorems near threshold and of QCD calculations at high energies. In the resonance region a better understanding of the $\pi\pi$ and K$\overline{\text{K}}$ continuum is needed to improve the analysis of resonances. Especially interesting is the study of the threshold behaviour of $\pi\pi$ and K$\overline{\text{K}}$ production because deviations from the Born cross sections could signal the presence of scalar resonances. Presently available results are inconclusive.

The measured cross section for the sum of $\pi^+\pi^-$, K^+K^- and p$\overline{\text{p}}$ production above 2 GeV is roughly of the same order of magnitude as predicted by perturbative QCD calculations. However, stringent QCD tests, including measurements of the energy dependence and of the angular distributions, need higher energies and better statistics. This is difficult to achieve because the expected rates drop fast with increasing energy.

The differential cross section for p$\overline{\text{p}}$ production has been measured up to p$\overline{\text{p}}$ masses of 3 GeV. In contrast to expectations from perturbative QCD, preliminary results indicate that $\Delta^{++}\overline{\Delta^{++}}$ production is smaller than p$\overline{\text{p}}$ production. A quantitative comparison of the measured baryon pair production with absolute theoretical predictions is at the moment not possible because the cross sections predicted by the two calculations available differ by two orders of magnitude in their relative normalization.

A large amount of experimental results is now available on TWO-PHOTON COUPLINGS OF RESONANCES. The $\gamma\gamma$ widths have been measured for the pseudoscalar (π°, η, η') and the tensor mesons (f, A$_2$, f'). The ratios of the measured couplings of the tensor mesons are consistent with the SU(3) model (assuming nonet symmetry). The SU(3) mixing angle is found to be close to the value for ideal mixing, in agreement with the GMO mass formula. The couplings of the pseudoscalars have still some experimental uncertainties. For the storage ring measurement of the $\gamma\gamma$ width of the η better statistics is needed to compare with the Primakoff measurement. The measured $\gamma\gamma$ width of the η' may systematically become smaller if the correct matrix element is used in the analysis. The SU(3) nonet mixing angle, determined from different combinations of $\gamma\gamma$ widths of the pseudoscalars, covers a range which includes the value obtained from the GMO mass formula. Besides the experimental uncertainties for the pseudoscalar couplings, which have still to be settled, there may be theoretical uncertainties in using the naive formulae for the mixing angles.

With the measured $\gamma\gamma$ width of the η' and some additional assumptions, the Han-Nambu model of integrally charged quarks is ruled out. However, these additional assumptions are not generally accepted.

The observed mass shift of the f resonance in the $\pi^+\pi^-$ channel seems to be naturally explained by interference with the continuum. It has not yet been shown conclusively that the $\pi^0\pi^0$ channel conforms with this interpretation. The investigation of the interference structure in the K$\overline{\text{K}}$ channel, where at least three resonances, f, A$_2$ and f', contribute with similar magnitude, is probably more difficult. An improvement in the measurement of the $\gamma\gamma$ width of the f' requires a better understanding of this channel. Close to threshold the K$\overline{\text{K}}$ channel may provide some information on scalar resonances.

From the analysis of $\gamma\gamma$ widths no indication for gluonium admixtures in "normal" mesons has been found. Upper limits for the two-photon couplings of the glueball candidates $\iota(1440)$ and $\Theta(1640)$ have been given, but for most of the models they are not stringent enough.

Little experimental information exists on the Q^2 dependence of the $\gamma\gamma$ widths of the mesons, which would allow testing quark wave functions. Whereas in the case of pseudoscalars only one form factor has to be measured, for tensor mesons and axial vectors (axial vectors can be produced by virtual photons) a complicated helicity structure has to be disentangled.

The cross section for $\rho^0\rho^0$ production shows a large enhancement near threshold. A spin-parity analysis indicates a dominance of $J^P=0^+$ and 2^+ states. The enhancement is not due to a single non-exotic resonance because the corresponding enhancement in the $\rho^+\rho^-$ channel, required by isospin symmetry, has not been observed. Production of exotic resonances, such as four-quark states, may give a possible explanation. Further studies of vector meson pair production near threshold (combinations of ρ, ω, φ, J/Ψ, K^*,...) should help to solve this puzzle.

The narrow structure observed in the four pion invariant mass around 2.1 GeV needs confirmation.

In the following we list some of the negative results of resonance searches:

- Two-photon production of scalar mesons has not been observed. Upper limits have been given for the S* and ε mesons. One should search for the scalars around 1 GeV in the high statistics data on $\gamma\gamma\to\pi^+\pi^-$ now available at PETRA and PEP.
- The search for two-photon production of the glueball candidates ι and Θ has been negative. Measurements of the $\gamma\gamma$ widths of these states would help to decide whether they are glueballs or not.
- Two-photon production of heavy quark states has not been observed. Measurements of the $\gamma\gamma$ widths of the charmonium states may be possible with higher integrated luminosity at PETRA and PEP.

The TOTAL CROSS SECTION for hadron production by two quasi-real photons has been studied. The experimental analysis is complicated because the detector acceptances depend strongly on the models used for the generation of the hadronic final state. More detailed studies of the sensitivity of the experiments to the models have to be carried out.

The two-photon production of hadronic final states at low Q^2 and not too large hadron transverse momenta shows similarities to hadronic reactions. The VMD-Regge estimates, which relate the two-photon total cross section to photon-hadron and hadron-hadron interactions, appear to give a reasonable description of the total cross section above the resonance region.

A measurement of the total cross section in the resonance region would test the Regge duality picture relating the t-channel exchanges at high energies to s-channel resonances. In the case of two-photon scattering an additional, anomalous contribution $\sim 1/W_{\gamma\gamma}^2$ is predicted from the pointlike coupling of the photon, which would result in a large cross section in the threshold region. The resonances known up to now cannot account for such a term.

The determination of the total cross section in the resonance region needs more experimental effort. Using an averaging, statistical model may not be adequate to describe the hadronic final states in this region. The model dependence in the simulation of the final state would be greatly reduced by a double-tag experiment with sufficient $W_{\gamma\gamma}$ resolution. It may also be possible to determine the total cross section in the low energy region by summing up all explicitly measured final states, i.e. completing Fig.7.2.

HARD SCATTERING PROCESSES in two-photon reactions offer the opportunity to study the QCD structure of hadron dynamics. The analyses of hard scattering processes in two-photon reactions show that the photon is different from a hadron, although in soft scattering processes the photon appears to interact like a hadron. The pointlike nature of the photon manifests itself in the measured transverse momentum distribution of hadrons produced by two photons. For transverse momenta above about 1 GeV this distribution is much flatter than those obtained in hadronic reactions at comparable energies. This different behaviour means that in two-photon reactions pointlike processes are more prominent than expected from a purely hadronic behaviour of the photon.

For a comparison with the Born diagram for $\gamma\gamma \to q\bar{q}$, the data have been analysed in terms of two-jet topologies. In single-tag data ($Q^2 \approx 0.3$ GeV^2) the cross section for the production of jets with a transverse momentum above 2 GeV was found to be about a factor of 2 larger than expected from the Born term calculated for fractionally charged quarks. Recent analyses indicate that the difference becomes larger for smaller Q^2 (quasi-real photons) and smaller for larger Q^2. The data are in agreement with the Born prediction for fractionally charged quarks for $Q^2 > 10$ GeV^2, irrespective of the transverse momentum of the jet. This rules out the Han-Nambu model of integrally charged quarks, while the gauge invariant version of this model, which predicts a Q^2 dependence of the effective quark charges, is still possible. The Q^2 dependence observed for the transverse momentum distribution of the hadrons or jets can also arise from higher order processes and higher twist contributions. The separation of the Born process from other contributions may be easier to achieve at higher energies (LEP, TRISTAN). However, a really stringent test of the integrally charged quark model probably requires a 0° tagging experiment so that the photons could be restricted to very small Q^2 values.

The first measurements of the STRUCTURE FUNCTION F_2 of the photon have given evidence for the pointlike behaviour of the photon when probed with a highly virtual photon. The observed characteristic features are: large values of F_2 for large x and an apparent increase of the structure function with Q^2. The absolute size of the transverse structure function and the increase with Q^2 depend, according to perturbative QCD calculations, on the QCD scale parameter Λ. For x values above ~ 0.3, the measurements are in good agreement with leading and next-to-leading order QCD calculations for Λ values $\Lambda_{LO} \approx \Lambda_{\overline{MS}} \approx 0.1...0.35$ GeV provided one adds a hadronic piece estimated by using the vector meson dominance model. On the other hand, the data are equally well described by using a simple quark-parton model instead of the QCD calculations, i.e. the measurements are no proof yet that QCD is needed.

The experimental investigation of the photon structure functions has made rapid progress in the last few years. Recent studies indicate that the unfolding of the x distribution can be reliably done for detectors with good angular acceptance for charged and neutral particles.

At the moment the real problem seems to lie in the theoretical basis for the Λ determination. The well known problems with the higher order QCD calculations around x=0 and x=1 are expanding to the range of medium x values, so that the convergence of the perturbative expansion appears not to be guaranteed at any x value. Different ways to circumvent these problems have been suggested, all of which imply that one has to give up the ambitious goal to determine Λ from an absolute, parameter-free measurement of the structure function. One suggestion is to introduce a new parameter which allows the regularization of unwanted singularities. Another is to determine Λ from the Q^2 evolution of the structure function.

In both cases, additional quantities have to be determined from the data, which requires a high precision determination of the x and Q^2 dependence of the structure functions. An extension to higher machine energies would increase the lever arm for a determination of the logarithmic Q^2 dependence of the transverse structure function.

In the future, the structure functions F_1 and F_3 may also be measured. The measurement of F_3 requires a double-tag experiment. In such an experiment one can also measure the structure functions of the virtual photon, allowing tests of QCD predictions which are much safer than in the case of quasi-real photons. Preliminary results on structure function measurements in a double-tag experiment have already been reported.

The large amount of experimental results on two-photon physics obtained in the last few years has greatly improved our understanding of the structure of photons and hadrons. The many questions, which had to be left open in this review, will hopefully be answered by the experiments to be completed in the near future.

References

/1/   H.Euler and B.Kockel, Nat.Wiss. 23 (1935) 246;
H.Euler, Ann.Physik 26 (1936) 398;
A.I.Akhiezer, Phys.Zeits.Sow. 11 (1937) 263;
R.Karplus. and M.Neuman, Phys.Rev. 83 (1951) 776;
Reviewed by: V.Costantini, B.De Tollis and G.Pistoni, Nuovo Cim. 2A (1971) 733

/2/   F.Low, Phys.Rev. 120 (1960) 582

/3/   F.Calogero and C.Zemach, Phys.Rev 120 (1960) 1860

/4/   A.Jaccarini, N.Arteaga-Romero, J.Parisi and P.Kessler, Compt.Rend. 269B (1969) 153, 1129; Nuovo Cim.4 (1970) 933;
V.E.Balakin, V.M.Budnev and I.F.Ginzburg, Zh.E.T.F.Pis'ma 11 (1970) 559 (JETP Lett.II, 388);
S.J.Brodsky, T.Kinoshita and H.Terazawa, Phys.Rev. D4 (1971) 1532;
H.Terazawa, Rev.Mod.Phys. 45 (1973) 615;
G.Bonneau, M.Gourdin and F.Martin, Nucl.Phys. B54 (1973) 573;
V.M.Budnev, I.F.Ginzburg, G.V.Meledin and V.G.Serbo, Phys.Rep. 15 (1975) 181.

/5/   V.E.Balakin et al., Phys.Lett. 34B (1971) 99

/6/   C.Bacci et al., Nuovo Cim.Lett. 3 (1972) 709

/7/   S.Brodsky, T.A. DeGrand, J.F. Gunion and J.H.Weis, Phys.Rev. D19 (1979) 1418

/8/   E.Witten, Nucl.Phys. B120 (1977) 189

/9/   W.Wagner, Proceedings of the XXth Intern. Conference on High Energy Physics, Madison, Wisconsin (1980), ed. L.Durand and L.G.Pondrom;
Ch.Berger, Proceedings of the EPS Intern. Conference on High Energy Physics, Lisbon (1981), ed. J.Dias de Deus and J.Soffer;
R.Wedemeyer, Proceedings of the 10th Intern. Symposium on Lepton and Photon Interactions at High Energies, Bonn (1981), ed. W.Pfeil;
S.Cooper, Proceedings of the Intern. Conference on Physics in Collision, Stockholm (1982);
D.L.Burke, Proceedings of the XXIst Intern. Conference on High Energy Physics, Paris (1982), ed. P.Petiau and M.Porneuf, Journal de Physique, tome 43, Coll. C-3, Suppl.12;
J.Dainton, Proceedings of the EPS Intern. Conference on High Energy Physics, Brighton (1983), ed. J.Guy and C.Costain;
Ch.Berger, Proceedings of the 11th Intern. Symposium on Lepton and Photon Interactions at High Energies, Ithaca (1983).

/10/ Proceedings of the Intern. Colloquium on Photon-Photon Collisions in Electron-Positron Storage Rings, Paris (1973), Journal de Physique, tome 35, Coll. C-2, Suppl.3;
Proceedings of the Intern. Conference on Two-Photon Interactions, Lake Tahoe, Calif. (1979), ed. J.F.Gunion;
Proceedings of the Intern. Workshop on $\gamma\gamma$-Collisions, Amiens, ed. G.Cochard and P.Kessler, Lecture Notes in Physics Vol.134, Springer Verlag (1980);
Proceedings of the 4th Intern. Colloquium on $\gamma\gamma$-Interactions, Paris (1981), ed. G.W.London;
Proceedings of the 5th Intern. Colloquium on $\gamma\gamma$-Interactions, Aachen, ed. Ch.Berger, Lecture Notes in Physics Vol.191, Springer Verlag (1983).

/11/ J.A.M.Vermaseren, Proceedings of the Intern. Workshop on $\gamma\gamma$-Collisions, Amiens, ed. G.Cochard and P.Kessler, Lecture Notes in Physics Vol.134, Springer Verlag (1980);
N.Arteaga-Romero, A.Jaccarini, P.Kessler and J.Parisi, Phys.Rev. D3 (1971) 1569;
J.Parisi, Proceedings of the Intern. Colloquium on Photon-Photon Collisions in Electron-Positron Storage Rings, Paris (1973), Journal de Physique, tome 35, Coll. C-2, Suppl.3;
R.M.Godbole and J.Smith, Nucl.Phys. B158 (1979) 234;
Ch.C.Peterson, T.F.Walsh and P.M.Zerwas, Nucl.Phys. B174 (1980) 424.

/12/ V.M.Budnev, I.F.Ginzburg, G.V.Meledin and V.G.Serbo, Phys.Rep. 15 (1975) 181

/13/ G.Bonneau, M.Gourdin and F.Martin, Nucl.Phys. B54 (1973) 573

/14/ G.Köpp, T.Walsh and P.Zerwas, Nucl.Phys. B70 (1974) 461

/15/ J.Field, Nucl.Phys. B168 (1980) 477 and Erratum B176 (1980) 545.

/16/ P.Kessler,Nuovo Cim. 17 (1960) 809

/17/ Ch.Berger and J.H.Field, Nucl.Phys. B187 (1981) 585

/18/ PLUTO Collaboration, PETRA Proposal PRC 79/06 and addendum;
Ch.Berger, Proc. of the Int. Conf. on Experimentation at LEP (Uppsala 1980), Phys.Scr. 23 (1981) 579.

/19/ D.Caldwell et al., PEP Proposal 9, Proposal for PEP Forward Detector Facility (1976)

/20/ H.J.Besch et al., Phys.Lett. 81B (1979) 79

/21/ A.Courau et al., Phys.Lett. 96B (1980) 402

/22/ ARGUS Collaboration, DESY proposal PRC 83/06 (1983)

/23/ G.Barbiellini et al., 1974 PEP Summer Study, PEP-B7, p.534

/24/ P.Jenni et al., Phys.Rev. D27 (1983) 1031

/25/ M.Oreglia et al., Phys.Rev. D25 (1982) 2259

/26/ C.J.Biddick et al., Phys.Lett. 97B (1980) 320

/27/ CELLO Collaboration, H.-J.Behrend et al., Phys.Scripta 23 (1981) 610

/28/ JADE Collaboration, W.Bartel et al., Phys.Lett. 88B (1979) 171

/29/ Mark J Collaboration, D.P.Barber et al., Phys.Rep. 63 (1980) 337

/30/ Ch.Berger, Proceedings of the Intern. Workshop on $\gamma\gamma$-Collisions, Amiens, ed. G.Cochard and P.Kessler, Lecture Notes in Physics Vol.134, Springer Verlag (1980)

/31/   TASSO Collaboration, R.Brandelik et al., Phys.Lett. 83B (1979) 261;
        Phys.Lett. 108B (1982) 71; Z.Phys. C10 (1981) 117
/32/   Christiaan Huygens, Traité de la Lumière, Leiden 1690
/33/   K.O.Mikaelian, Phys.Lett. 115B (1982) 267
/34/   S.I.Vavilov, Jour.Russ.Phys.Chem. 60 (1928) 555; Phys.Rev. 36 (1930)
        1590;
        A.L.Hughes and G.E.M.Jauncey, Phys.Rev. 36 (1930) 773.
/35/   A.Weinstein et al., Phys.Rev. D28 (1983) 2896
/36/   S.J.Brodsky, T.Kinoshita and H.Terazawa, Phys.Rev. D4 (1971) 1532
/37/   L.D.Landau and E.M.Lifshitz, Sov.Phys. 6 (1934) 320
/38/   J. Smith, J.A.M.Vermaseren, G. Grammer, Phys. Rev. D15 (1977) 3280;
        J.A.M.Vermaseren, Proceedings of the Intern. Workshop on
        $\gamma\gamma$-Collisions, Amiens, ed. G.Cochard and P.Kessler, Lecture Notes in
        Physics Vol.134, Springer Verlag (1980);
        J.A.M.Vermaseren, Program write-up, unpublished.
/39/   F.A.Berends, P.H.Daverveldt and R.Kleiss, contributed paper to the
        5th Intern. Colloquium, Lecture Notes in Physics Vol.191, Springer
        Verlag (1983).
/40/   G.Barbiellini, Proceedings of the Intern. Workshop on $\gamma\gamma$-Collisions,
        Amiens, ed. G.Cochard and P.Kessler, Lecture Notes in Physics
        Vol.134, Springer Verlag (1980)
/41/   A.Courau et al., Phys.Lett. 96B (1980) 402;
        R.Wedemeyer, Proceedings of the 10th Intern. Symposium on Lepton
        and Photon Interactions at High Energies, Bonn (1981), ed. W.Pfeil
        and paper No.48 submitted to the conference
/42/   Mark J Collaboration, B.Adeva et al., Phys.Rev.Lett. 48 (1982) 721
/43/   PLUTO Collaboration, Ch. Berger et al., Phys. Lett. 94B (1980) 254
/44/   TASSO Collaboration, R.Brandelik et al., Z.Phys. C10 (1981) 117
/45/   A.Roussarie et al., Phys. Lett. 105B (1981) 304
/46/   PLUTO Collaboration, Ch.Berger et al., Nucl.Phys. B202 (1982) 189
/47/   CELLO Collaboration, H.-J.Behrend et al., Phys.Lett. 126B (1983) 384
/48/   G.Cochard and S.Ong, Phys.Rev. D19 (1979) 810;
        M.Defrise, S.Ong, J.Silva and J.Carimalo, Phys.Rev. D23 (1981) 663;
        M.Defrise, Z.Phys. C9 (1981) 41;
        Y.Srivastava, Proceedings of the 4th Intern. Colloquium on
        $\gamma\gamma$-Interactions, Paris (1981), ed. G.W.London;
        F.A.Berends, P.H.Daverveldt and R.Kleiss, Leiden University preprint
        (1983);
        W.L.van Neerven and J.A.M.Vermaseren, NIKHEF-H/83-11 (1983).
/49/   D.H.Lyth, Nucl.Phys. B30 (1971) 195;
        G.Schierholz and K.Sundermeyer, Nucl.Phys. B40 (1972) 125;
        O.Babelon et al., Nucl.Phys. B113 (1976) 445.
/50/   G.Mennessier, Z.Phys. C16 (1983) 241
/51/   K.M.Watson, Phys.Rev. 95 (1954) 228
/52/   G.P.Lepage and S.J.Brodsky, Phys.Rev. D22 (1980) 2157;
        S.J.Brodsky and G.P.Lepage, Phys.Rev. D24 (1981) 1808.

/53/   H.Kolanoski, Proceedings of the 5th Intern. Colloquium on γγ-Interactions, Aachen, ed. Ch.Berger, Lecture Notes in Physics Vol.191, Springer Verlag (1983)

/54/   TASSO Collaboration, R.Brandelik et al., Phys. Lett. 108B (1982) 67

/55/   TASSO Collaboration, M.Althoff et al., Phys.Lett. 130B (1983) 449

/56/   N.Arteaga-Romero, Seminar on γγ-Physics, LPC/82-14, Paris (1982)

/57/   CERN-HERA 79-03 data compilation

/58/   P.H.Damgaard, Nucl.Phys. B211 (1983) 435

/59/   G.R.Farrar, E.Maina and F.Neri, Rutgers University RU-83-33

/60/   P.H.Damgaard, private communication

/61/   TASSO Collaboration, M.Althoff et al., to be published

/62/   G.S.Abrams et al., Phys.Rev.Lett. 43 (1979) 477

/63/   C.N.Yang, Phys.Rev. 77 (1950) 242

/64/   H.Fritzsch and P.Minkowski, Nuovo Cim. 30A (1975) 393;
        R.P.Freund and Y.Nambu, Phys.Rev.Lett. 34 (1975) 1645;
        J.F.Bolzan, W.F.Palmer and S.S.Pinsky, Phys.Rev. D14 (1976) 3202.

/65/   R.Waldi, K.R.Schubert and K.Winter, Z.Phys. C18 (1983) 301

/66/   R.L.Jaffe, Phys.Rev. D15 (1977) 267 and 281;
        R.L.Jaffe and K.Johnson, Phys.Lett. 60B (1976) 201.

/67/   L.Montanet, Rep.Prog.Phys. 46 (1983) 337

/68/   T.Appelquist R.M.Barnet and K.D.Lane, Ann. Rev. Nucl. Part.Sci. 28 (1978) 387;
        C.Quigg and J.L.Rosner, Phys.Rep. 56 (1979) 167.

/69/   F.-M.Renard, Electron-Positron Collisions, Edition Frontière, Gif sur Yvette, France, 1981

/70/   A.DeRujula, H.Georgi and S.L.Glashow, Phys.Rev. D12 (1975) 147;
        J.D.Jackson, Proc. of the Summer Institute on Particle Physics, Stanford 1976, SLAC Report No.198.

/71/   S.L.Adler, Phys.Rev. 177 (1969) 2426;
        J.S.Bell and L.Jackiw, Nuovo Cimento 60A (1969) 47;
        S.L.Adler, Lectures on Elementary Particles and Quantum Field Theory, Vol.1, Brandeis 1970, MIT Press, Cambridge (1971).

/72/   Particle Data Group, Phys.Lett. 111B (1982) 1

/73/   C.Bemporad et al., Phys.Lett. 25B (1967) 380

/74/   A.Browman et al., Phys.Rev.Lett. 32 (1974) 1067

/75/   L.Paoluzi et al., Lett.Nuovo Cim. 10 (1976) 435

/76/   D.M.Binnie et al., Phys.Lett. 83B (1979) 141

/77/   JADE Collaboration, W.Bartel et al., Phys.Lett. 113B (1982) 190

/78/   CELLO Collaboration, H.-J.Behrend et al., Phys.Lett. 114B (1982) 378 and Erratum, Phys.Lett. 125B (1983) 518

/79/   A.Rittenberg, Thesis, UCRL-18863, Berkeley 1969 (unpublished)

/80/   J.E.Olsson, Proceedings of the 5th Intern. Colloquium on γγ-Interactions, Aachen, ed. Ch.Berger, Lecture Notes in Physics Vol.191, Springer Verlag (1983)

/81/   J.Dainton, Proceedings of the EPS Intern. Conference on High Energy Physics, Brighton (1983), ed. J.Guy and C.Costain

/82/   M.Poppe, Proceedings of the XIV. International Symposium on Multiparticle Dynamics, Granlibakken, Lake Tahoe, USA (1983), (preprint DESY 83-088)

/83/ K.Wilson, Phys.Rev. 179 (1969) 1499

/84/ S.Weinberg, Phys.Rev. D11 (1975) 3583;
H.Fritzsch, M.Gell-Mann and H.Leutwyler, Phys.Lett. 47B (1973) 365;
G.'t Hooft, Phys.Rev.Lett. 37 (1976) 8; Phys.Rev. D14 (1976) 3432.

/85/ H.Suura, T.F.Walsh and B.-L.Young, Lett.Nuovo Cim. 4 (1972) 505 ess,

/86/ M.Y.Han and Y.Nambu, Phys.Rev. 139 (1965) B1006

/87/ M.S.Chanowitz, Phys.Rev.Lett. 44 (1980) 59

/88/ F.J.Gilman, Proceedings of the Intern. Conference on Two-Photon Interactions, Lake Tahoe, Calif. (1979), ed. J.F.Gunion

/89/ J.H.Field, Proceedings of the EPS Intern. Conference on High Energy Physics, Brighton (1983), ed. J.Guy and C.Costain

/90/ C.Edwards et al., Phys.Rev.Lett. 49 (1982) 259

/91/ E.Bloom, XXI International Conference on High Energy Physics, Paris (1982)

/92/ K.Ishikawa, Phys.Rev.Lett. 46 (1981) 978.

/93/ H.Goldberg, Phys.Rev. D22 (1980) 2286

/94/ S.Pinsky, Proc. of the XVIIth Rencontre de Moriond, Les Arcs, France, March 20-26 1982

/95/ F.Renard, Proceedings of the 5th Intern. Colloquium on $\gamma\gamma$-Interactions, Aachen, ed. Ch.Berger, Lecture Notes in Physics Vol.191, Springer Verlag (1983);

/96/ G.M.Radutskij, Sov.Jour.Nucl.Phys. 8 (1969) 65

/97/ B.Schrempp-Otto, F.Schrempp and T.Walsh, Phys.Lett. 36B (1971) 463

/98/ B.Renner, Nucl.Phys. B30 (1971) 634

/99/ A.Bramon and M.Greco, Lett.Nuovo Cim. 2 (1971) 522

/100/ G.Schierholz and K.Sundermeyer, Nucl.Phys. B40 (1972) 125

/101/ V.N.Novikov and S.I.Eidelmann, Sov.Jour.Nucl.Phys. 21 (1969) 65

/102/ J.Babcock and J.L.Rosner, Phys.Rev. D14 (1976) 1286

/103/ P.Grassberger and R.Kögerler, Nucl.Phys. B106 (1976) 451

/104/ V.M.Budnev and A.E.Kaloshin, Phys.Lett. 86B (1979) 351

/105/ P.Singer, Phys.Lett. 124B (1983) 531

/106/ H.Krasemann and J.A.M.Vermaseren, Nucl.Phys. B184 (1981) 269

/107/ S.J.Brodsky and G.P.Lepage, Proceedings of the XXth Intern. Conference on High Energy Physics, Madison, Wisconsin (1980), ed. L.Durand and L.G.Pondrom

/108/ C.Edwards et al., Phys.Lett. 110B (1982) 82

/109/ J.Haissinski, Proc. of the XVIIth Rencontre de Moriond, Les Arcs, France, March 20-26 1982

/110/ TASSO Collaboration, M.Althoff et al., Phys.Lett. 121B (1983) 216

/111/ H.J.Lipkin, Nucl.Phys. B7 (1968) 321; Proceedings of the EPS Intern. Conference on High Energy Physics, Palermo (1975);
D.Faiman, H.J.Lipkin and H.R.Rubinstein, Phys.Lett. 59B (1975) 269.

/112/ C.Edwards et al., Phys.Rev.Lett. 48 (1982) 458

/113/ D.L.Burke et al., Phys.Rev.Lett. 49 (1982) 632

/114/ TASSO Collaboration, M.Althoff et al., Z.Phys. C16 (1982) 13

/115/ K.Wacker, Proc. of the XVIIIth Rencontre de Moriond, La Plagne, France, March 13-19, 1983

/116/ D.L.Burke, Proceedings of the XXIst Intern. Conference on High Energy Physics, Paris (1982), ed. P.Petiau and M.Porneuf, Journal de Physique, tome 43, Coll. C-3, Suppl.12

/117/ H.J.Schnitzer, Nucl.Phys B207 (1982) 131;
J.L.Rosner, Phys.Rev. D24 (1981) 1347;
J.L.Rosner, Phys.Rev. D27 (1983) 1101;
M.Nicolič, Phys.Rev D26 (1982) 3020;
T.Teshima and S.Oneda, Phys.Rev. D27 (1983) 1551.

/118/ J.L.Rosner and S.F.Tuan, University of Hawaii preprint, UH-511-477-82 (1982)

/119/ A.B.Wicklund et al., Phys.Rev.Lett. 45 (1980) 1469

/120/ J.Babcock and J.L.Rosner, Phys.Rev. D14 (1976) 1286

/121/ J.Crewther, Phys.Rev.Lett 28 (1972) 1421;
M.S.Chanowitz and J.Ellis, Phys.Rev. D7 (1973) 2490.

/122/ M.Greco, Proceedings of the Intern. Workshop on $\gamma\gamma$-Collisions, Amiens, ed. G.Cochard and P.Kessler, Lecture Notes in Physics Vol.134, Springer Verlag (1980)

/123/ P.Roy, Phys.Rev. 9 (1974) 2631

/124/ M.Greco and Y.Srivastava, Nuovo Cim. 43A (1978) 88

/125/ V.A.Novikov et al., Phys.Rep. 41 (1978) 1;
L.J.Reinders, H.R.Rubinstein and S.Yazaki, Nucl.Phys. B186 (1981) 109.

/126/ M.A.Shifman, A.I.Vainshtein, M.B.Voloshin and V.I.Zakharov, Phys.Lett. 77B (1978) 80;
R.Kirschner and A.Schiller, Z.Phys. C16 (1982) 141;
T.M.Aliev, Yad.Fiz. 37 (1983) 403 (ITEF-82-76 preprint, Moscow).

/127/ L.J.Reinders, H.R.Rubinstein and S.Yazaki, Phys.Lett. 113B (1982) 411

/128/ M.Shifman and M.Voysotsky, Z.Phys. C10 (1981) 131

/129/ E.D.Bloom and C.W.Peck, Ann.Rev.Nucl.Part.Sci. 33 (1983) 143

/130/ R.Partridge et al., Phys.Rev.Lett. 44 (1980) 712

/131/ T.M.Himel et al., Phys.Rev.Lett. 45 (1980) 1146

/132/ R.Partridge et al., Phys.Rev.Lett. 45 (1980) 1150

/133/ K.F.Einsweiler, Proceedings of the EPS Intern. Conference on High Energy Physics, Brighton (1983), ed. J.Guy and C.Costain

/134/ M.Oreglia, Proc. of the XVth Rencontre de Moriond, Les Arcs, France, March 15-21, 1980

/135/ J.E.Gaiser, Proc. of the XVIIth Rencontre de Moriond, Les Arcs, France, Jan. 24-30, 1982

/136/ TASSO Collaboration, R.Brandelik et al., Phys.Lett. 97B (1980) 448

/137/ D.L.Burke et.al., Phys. Lett. 103B (1981) 153

/138/ CELLO Collaboration, H.-J.Behrend et al., Z.Phys. C21 (1984) 205

/139/ G.Alexander, U.Maor and P.G.Williams, Phys.Rev. D26 (1982) 1198

/140/ J.Layssac and F.M.Renard, Montpelier preprint PM/80/11 (1980)

/141/ H.Goldberg and T.Weiler, Phys. Lett. 102B (1981) 63

/142/ R.M.Godbole and K.V.L.Sarma, Phys. Lett. 109B (1982) 504

/143/ S.Minami, Lett. Nuov. Cim. 34 (1982) 125

/144/ Bing An Li and K.F.Liu, Phys.Lett. 118B (1982) 435 and Erratum, Phys.Lett. 124B (1983) 550

/145/ N.N.Achasov, S.A.Devyanin, and G.N.Shestakov, Phys. Lett. 108B (1982) 134; Z.Phys. C16 (1982) 55.

/146/ K.Biswal and S.P.Misra, Phys.Rev. D26 (1982) 3020

/147/ Particle Data Group, C.Bricman et al., Phys. Lett. 75B (1978)

/148/ J.H.Field, Proceedings of the 4th Intern. Colloquium on γγ-Interactions, Paris (1981), ed. G.W.London

/149/ M.Krammer, private communication

/150/ N.N.Achasov, private communication

/151/ M.S.Chanowitz, Lectures given at the SLAC Summer Institute, Stanford, 1981

/152/ D.Lüke, Proceedings of the XXIst Intern. Conference on High Energy Physics, Paris (1982), ed. P.Petiau and M.Porneuf, Journal de Physique, tome 43, Coll. C-3, Suppl.12;
       H.Kolanoski, Proc. of the Seminar on γγ Physics, Montpellier, Dec. 9-10, 1982, University of Montpellier PM 83/6 (1983).

/153/ J.L.Rosner, Brookhaven report CRISP 71.26 (1971)

/154/ S.J.Brodsky, F.E.Close and J.F.Gunion, Phys.Rev.D6 (1972) 177

/155/ J.F.Gunion, Proceedings of the Intern. Workshop on γγ-Collisions, Amiens, ed. G.Cochard and P.Kessler, Lecture Notes in Physics Vol.134, Springer Verlag (1980)

/156/ R.P.Worden, Phys.Lett. 52B (1974) 87

/157/ D.L.Burke, Proceedings of the Intern. Workshop on γγ-Collisions, Amiens, ed. G.Cochard and P.Kessler, Lecture Notes in Physics Vol.134, Springer Verlag (1980)

/158/ PLUTO Collaboration, Ch.Berger et al., Phys.Lett. 89B (1981) 287;
       F.A.Raupach, Thesis, DESY PLUTO 81/10 (1981) (unpublished).

/159/ E.Hilger, Proceedings of the Intern. Workshop on γγ-Collisions, Amiens, ed. G.Cochard and P.Kessler, Lecture Notes in Physics Vol.134, Springer Verlag (1980)

/160/ W.Hillen, Thesis Bonn 1981, BONN-IR-81-7 (unpublished)

/161/ I.F Ginzburg and V.G.Serbo, Phys.Lett. 109B (1982) 231

/162/ J.H.Friedman, Journ.Comp.Phys. 7 (1971) 201;
       O.Pene and A.Krzywicki, Nucl.Phys. B12 (1969) 415.

/163/ G.G.Hanson et al., Phys.Rev.Lett. 35 (1975) 1609;
       S.Cooper, LBL-Report 11322 (1980), (Thesis, unpublished);
       J.L.Siegrist et al., Phys.Rev. D26 (1982) 969, 991.

/164/ Z.Koba, H.B.Nielsen and P.Olesen, Nucl.Phys. B40 (1980) 313

/165/ W.Wagner, Proceedings of the 5th Intern. Colloquium on γγ-Interactions, Aachen, ed. Ch.Berger, Lecture Notes in Physics Vol.191, Springer Verlag (1983)

/166/ Ch.Berger, Proceedings of the 4th Intern. Colloquium on γγ-Interactions, Paris (1981), ed. G.W.London

/167/ W.Wagner, private communication

/168/ N.Wermes, Thesis Bonn 1982, BONN-IR-82-27 (unpublished)

/169/ TASSO Collaboration, R.Brandelik et al., Phys.Lett. 89B (1980) 418

/170/ PLUTO Collaboration, C.Berger et al., Phys.Lett. 95B (1980) 313

/171/ S.M.Berman, J.D.Bjorken and J.B.Kogut, Phys.Rev. D4 (1971) 3388

/172/ J.C.Pati and A.Salam, Phys.Rev. D8 (1973) 1240; Phys.Rev. D10 (1974) 275.

/173/ K.Kajantie and R.Raitio, Nucl.Phys. B159 (1979) 528;
      K.Kajantie, Phys.Scripta 29 (1979) 230; Acta Phys.Austr.Suppl.XXI (1979) 663.

/174/ K.Kajantie, Proceedings of the 4th Intern. Colloquium on $\gamma\gamma$-Interactions, Paris (1981), ed. G.W.London

/175/ J.H.Field, E.Pietarinen and K.Kajantie, Nucl.Phys. B171 (1980) 377

/176/ J.A.Bagger and J.F.Gunion, University of California, Davis, Preprint UCD-83/1

/177/ C.H.Llewellyn Smith, Phys.Lett 79B (1978) 83

/178/ F.A.Berends, Z.Kunszt and R.Gastmans, Phys.Lett. 92B (1980) 186 and DESY Report 80/89 (1980)

/179/ G.Sterman and S.Weinberg, Phys.Rev.Lett. 39 (1977) 1436

/180/ TASSO Collaboration, R.Brandelik et al., Phys.Lett. 107B (1981) 290

/181/ I.Kang, Oxford Preprint 50/82 (1982)

/182/ J.Stirling, Proceedings of the 5th Intern. Colloquium on $\gamma\gamma$-Interactions, Aachen, ed. Ch.Berger, Lecture Notes in Physics Vol.191, Springer Verlag (1983)

/183/ F.Khalafi, P.V.Landshoff and W.J.Stirling, Phys.Lett. 130B (1983) 215

/184/ N.Wermes, Proceedings of the 5th Intern. Colloquium on $\gamma\gamma$-Interactions, Aachen, ed. Ch.Berger, Lecture Notes in Physics Vol.191, Springer Verlag (1983)

/185/ K.Lanius, DESY Report 80-36 (1980);
      H.J.Daum, H.Meyer and J.Bürger, Z.Phys. C8 (1981) 167.

/186/ D.Cords, Proceedings of the 4th Intern. Colloquium on $\gamma\gamma$-Interactions, Paris (1981), ed. G.W.London

/187/ R.D.Field and R.P.Feynman, Nucl.Phys. B137 (1978) 1

/188/ TASSO Collaboration, M.Althoff et al., DESY 83-115 (1983), submitted to Phys.Lett.

/189/ JADE Collaboration, W.Bartel et al., Phys.Lett. 107B (1981) 163

/190/ B.Alper et al., Nucl. Phys. B100 (1975) 237;
      for a review see: G.Giacomelli and M.Jacob, Phys.Rep. 55 (1979) 1.

/191/ A.Janah and M.Özer, University of Maryland, Preprint PUB-81-221;
      A.Janah, Doctoral dissertation, University of Maryland, 1982 (unpublished).

/192/ T.Jayarama et al., Phys.Lett. 119B (1982) 215;
      A.V.Efremov and S.V.Ivanov, Yad.Fiz. 35 (1982) 245;
      M.Katuya, Shizuoka Women's University, Japan, Prog.Theor.Phys. 68 (1982) 1802.

/193/ S.J.Brodsky, T.Kinoshita and H.Terazawa, Phys.Rev.Lett. 27 (1971) 280;
      T.F.Walsh, Phys.Lett. 36B (1971) 121.

/194/ T.F.Walsh and P.M.Zerwas, Phys.Lett. 44B (1973) 195;
R.L.Kingsley, Nucl.Phys. B60 (1973) 43;
R.P.Worden, Phys.Lett. 51B (1974) 57;
M.A.Ahmed and G.G.Ross, Phys.Lett. 59B (1975) 369.

/195/ W.Frazer, Proceedings of the 5th Intern. Colloquium on
$\gamma\gamma$-Interactions, Aachen, ed. Ch.Berger, Lecture Notes in Physics
Vol.191, Springer Verlag (1983)

/197/ D.W.Duke, Proceedings of the 5th Intern. Colloquium on
$\gamma\gamma$-Interactions, Aachen, ed. Ch.Berger, Lecture Notes in Physics
Vol.191, Springer Verlag (1983)

/198/ Ch.C.Petersen, T.F.Walsh and P.M.Zerwas, Nucl.Phys. B174 (1980) 424

/199/ K.J.Anderson et al., Chicago-Princeton Report EFI-78-38(1978)

/200/ G.Rossi, University of California, San Diego, USCD-10P10-227 (1983)

/201/ S.Cooper, Proceedings of the Intern. Conference on Physics in Collision, Stockholm (1982)

/202/ C.H.Llewellyn Smith, Phys.Lett. 79B (1978) 83;
DeWitt et al., Phys.Rev. D19 (1979) 2046;
W.R.Frazer and J.F.Gunion, Phys.Rev. D20 (1979) 147;
A.Nicolaidis, Nucl.Phys. B163 (1979) 156.

/203/ G.Altarelli and G.Parisi, Nucl.Phys. B126 (1977) 298

/204/ G.Altarelli, Proceedings of the Intern. Workshop on $\gamma\gamma$-Collisions,
Amiens, ed. G.Cochard and P.Kessler, Lecture Notes in Physics
Vol.134, Springer Verlag (1980)

/205/ W.A.Bardeen and A.J.Buras, Phys.Rev. D20 (1979) 166

/206/ S.J.Brodsky, SLAC-PUB-2447 (1979)

/207/ D.W.Duke and J.F.Owens, Phys.Rev. D22 (1980) 2280

/208/ I.Antoniadis and G.Grunberg, Nucl.Phys. B213 (1983) 445

/209/ G.Rossi, Phys.Lett. 130B (1983) 105

/210/ M.Glück and E.Reya, Phys.Rev. D28 (1983) 2749

/211/ W.A.Bardeen, private communication;
M.K.Chase, Nucl.Phys. B189 (1981) 461

/212/ T. Uematsu and T.F.Walsh, Phys.Lett. 101B (1981) 263

/213/ T. Uematsu and T.F.Walsh, Nucl.Phys. B199 (1982) 93

/214/ PLUTO Collaboration, Ch.Berger et al., Phys.Lett. 107B (1981) 168

/215/ JADE Collaboration, W.Bartel et al., Phys.Lett. 121B (1983) 216

/216/ CELLO Collaboration, J.-H.Behrend et al., Phys.Lett. 118B (1982) 211

/217/ T.Nozaki, private communication

/218/ V.Blobel, contribution to 5th Intern. Colloquium on $\gamma\gamma$-Interactions,
Aachen (1983),
reviewed by: H.Spitzer, Proceedings of the 5th Intern. Colloquium on
$\gamma\gamma$-Interactions, Aachen, ed. Ch.Berger, Lecture Notes in Physics
Vol.191, Springer Verlag (1983).

/219/ W.R.Frazer and J.F.Gunion, Phys.Rev. D20 (1979) 147

/220/ U.Maor and E.Gotsman, Phys.Rev. D28 (1983) 2149

/221/ CELLO Collaboration, J.-H.Behrend et al., Phys.Lett. 126B (1983) 391

Subject Index

G. Kramer

Theory of Jets in Electron-Positron Annihilation

1984. 86 figures. VII, 140 pages
(Springer Tracts in Modern Physics, Volume 102)
ISBN 3-540-10368-3

Contents: Introduction. – Eletron-Positron Annihilation into Hadron Jets. – e^+e^- Annihilation into Jets in QCD Perturbation Theory. – Summary and Conclusions. – References.

Quarks and Nuclear Forces

Editors: **D. C. Fries, B. Zeitnitz**
1982. 69 figures. XI, 223 pages
(Springer Tracts in Modern Physics, Volume 100)
ISBN 3-540-11717-2

Contents: *G. Flügge:* Experimental Evidence of Quarks and Gluons. – *F. E. Close:* The Pauli Principle and QCD for Quarks and Nucleons in Hadrons and Nuclei. – *S. J. Brodsky:* Quantum Chromodynamics at Nuclear Dimensions. – *L. Heller:* Bag Models and Nuclear Forces. – *G. Baym:* Quark Matter and Nuclei. – **A. Faessler:** How Should or Will QCD Infuence Nucelar Physics?

Modern Three-Hadron Physics

Editor: **A. W. Thomas**
1977. 30 figures. XI, 250 pages
(Topics in Current Physics, Volume 2)
ISBN 3-540-07950-5

Contents: *I. R. Afnan, A. W. Thomas:* Fundamentals of Three-Body Scattering Theory. – *L. R. Dodd:* Analytic Structure of On-Shell Three-Body Amplitudes. – *R. D. Amado:* Theory of Three-Body Final States. – *D. D. Brayshaw:* The Boundary Condition Method. – *R. Aaron:* A Relativistic Three-Body Theory. – *E. F. Redish:* Applications of Three-Body Methods to Many-Body Hadronic Systems.

Springer-Verlag
Berlin
Heidelberg
New York
Tokyo

Zeitschrift für Physic C

Particles and Fields

EPS Europhysics Journal

ISSN 0170-9739   Title No. 288

Editors in Chief: G. Kramer, Hamburg; H. Satz, Bielefeld

Editors: R. Barbieri, Pisa; T. Ferbel, Rochester; K. Fujikawa, Hiroshima; P. Hasenfratz, Genf; K. Kajantie, Helsinki; A. Krzywicki, Orsay; P. Söding, Hamburg; B. Stech, Heidelberg; J. C. Taylor, Cambridge; F. Wilczek, Santa Barbara

Zeitschrift für Physik C, **Particles and Fields** is devoted to the experimental and theoretical investigation of elementary particles. In view of the steadily growing interplay of theory and experiment in this field, particular emphasis is given to a clear and complete presentation of research.

The topics covered include:

- Experimental and theoretical particle physics
- Structure of elementary particles
- High energy processes
- Strong, electromagnetic and weak interactions
- Symmetry principles
- Unification schemes
- S-matrix theory
- Quantum field theory
- Lattice field theory

Special features: Rapid publication, no page charge. Language of publications is English.

Zeitschrift für Physik appears in three parts:
     A: Atoms and Nuclei
     B: Condensed Matter
     C: Particles and Fields

Each part may be ordered separately.

Coordinating editor for Zeitschrift für Physik, Parts A, B and C: O. Haxel, Heidelberg

Subscription information and/or sample copies are available from your bookseller or directly from Springer-Verlag, Journal Promotion Dept., P. O. Box 105 280, D-6900 Heidelberg, FRG

Springer-Verlag
Berlin
Heidelberg
New York
Tokyo